CAPSULE CALCULUS

Ira Ritow

Dover Publications, Inc.
Mineola, New York

Bibliographical Note

This Dover edition, first published in 2003, is an unabridged republi-
cation of the work originally published (as a Dolphin Book) by
Doubleday & Company, Inc., Garden City, New York, in 1962.

Library of Congress Cataloging-in-Publication Data

Ritow, Ira.
    Capsule calculus / Ira Ritow.
        p. cm.
    Originally published: Garden City, N.Y. : Doubleday, [1962]
    Includes index.
    ISBN 0-486-43256-4 (pbk.)
    1. Calculus. I. Title.

QA303.2.R58 2003
515—dc21

                                            2003053067

Manufactured in the United States of America
Dover Publications, Inc., 31 East 2nd Street, Mineola, N.Y. 11501

TO MY NATALIE

PREFACE

This book is an introduction to calculus—designed to present the principles in simple, easy-to-understand terms. As a result it can be used as an introductory survey course in the subject or as a review text. The material includes the basic principles ordinarily covered in two years of courses with enough detail to enable the reader to solve ordinary engineering and scientific problems requiring techniques of modern calculus.

The reader is assumed to have a mathematical background extending only through algebra and trigonometry and some familiarity with graphs. Additional background material is presented as needed.

The emphasis throughout is on developing the familiarity with basic principles usually called "understanding" rather than on presenting detailed proofs of accepted principles. Thus, the sequence of topics may seem unusual to readers nurtured on more conventional texts.

The origin of most of the material that follows was a series of articles I wrote for the magazine, Electrical Manufacturing (now Electro-Technology). I wish to thank C-M Technical Publications, the publishers of Electro-Technology, for permitting me to use material originally published by them, in *Capsule Calculus*.

Ira Ritow

TABLE OF CONTENTS

LIST OF ILLUSTRATIONS

I

DIFFERENTIAL CALCULUS

1. DIFFERENTIATION

DEFINITIONS: Assume that x and y are related numbers with some physical meaning. For example, x might be the distance driven by a car and y the amount of gas in the gas tank.

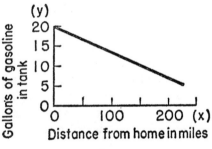

Fig. 1

As shown in Fig. 1, *for every value of x there exists a value of y*. This can also be stated as: "y is a function of x" [sometimes abbreviated as $y = f(x)$]. With but a single value of y for every value of x in this case, it can be said that "y is a *single valued* function of x." If for every value of x there were two or more possible values of y, the statement of their relationship would be: "y is a *multiple-valued* function of x." If y is a single-valued function of x, then x may or may not also be a single-valued function of y.

In plotting the relationship between distance and gasoline for a long trip, such as shown in Fig. 2, y is a single-valued function of x except at $x = 150$ and $x = 300$ where

y has two values. Such breaks in the graph are called "discontinuities" and it is said that y is a *discontinuous* function of x at $x = 150$ and at $x = 300$. Also, for some values of y (such as 15 gallons) x may have any one of several values. Therefore, in Fig. 2, x is a multiple-valued function of y.

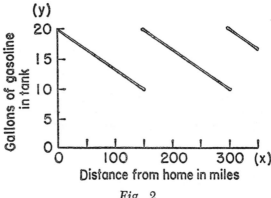

Fig. 2

Another useful term is the word "limit" (abbreviated "lim"). The limit value of y, if y is a function of x, is the value that y approaches as x approaches any given value. For example, in Fig. 2, as x approaches a value of 75 miles, the limit of y is 15 gallons. This can be written

$$\lim_{(x \to 75)} y = 15$$

But as $x \to 150$ in Fig. 2, $\lim y = 20$ or 10 depending on whether x approaches the value of 150 from the right or the left. Therefore the limit of a function at a discontinuity does not have a single value.

This book deals principally with single-valued continuous functions, such as shown in Fig. 1, where the limit value of y at any given value of x is a single value.

DIFFERENTIATION. The first derivative of a function is the slope of the graph of the function at every point. The slope of a graph at any point is a measure of its steepness at that point. As an example, examine Fig. 3.

The slope is fairly constant from $x = a$ to $x = a'$, a

short distance away, and can be expressed as the ratio of
the change in y to the change in x, or $\Delta y \div \Delta x$. Since
$\Delta y / \Delta x = 1$ at point a, the slope (or "first derivative of y with
respect to x") at $x = a$ is 1. Applying the same principle at
$x = e$ would give a negative value of Δy for a positive value

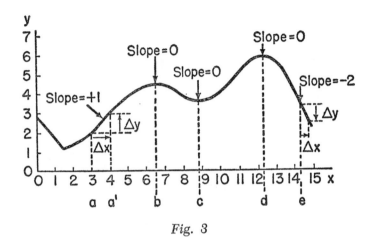

Fig. 3

of Δx. As a result, the ratio $\Delta y / \Delta x$ is negative at $x = e$. If
the first derivative of y with respect to x were calculated at
$x = a$ by taking a' to the left of a, then both Δy and Δx would
have been negative and the first derivative of y with respect
to x, or $\Delta y / \Delta x$, would have been a negative number divided
by a negative number or a positive number. At $x = b$, c, and
d, $\Delta y / \Delta x$ is zero divided by a finite number; so that the first
derivative of y with respect to x is zero at these points.

As the slope is not always constant over a large part of
the curve, Δx and Δy must often be taken as microscopically
small quantities to obtain even a fair approximation to the
slope at a point. In fact, for utmost accuracy Δx must be
almost zero. Expressed in mathematical shorthand, the basic
definition of the first derivative of y with respect to x is:

$$\frac{dy}{dx} = \lim_{(\Delta x \to 0)} \left[\frac{\Delta y}{\Delta x} \right]$$

This equation will now be applied to find the first derivative (dy/dx) of the function $y = x^2$ at the point $x = 2$, as shown in Fig. 4.

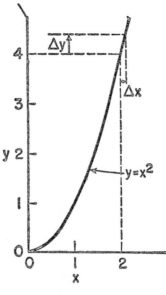

Fig. 4

At $x = 2$, $y = 4$. If x is increased by Δx, y becomes

$$y' = (x + \Delta x)^2 = (2 + \Delta x)^2$$

or

$$y' = 4 + 4\Delta x + (\Delta x)^2$$

In this case,

$$\Delta y = y' - y$$

so that

$$\Delta y = 4 + 4\Delta x + (\Delta x)^2 - 4$$
$$= 4\Delta x + (\Delta x)^2$$

Now

$$\frac{\Delta y}{\Delta x} = \frac{4\Delta x + (\Delta x)^2}{\Delta x} = \frac{4 + \Delta x}{1}$$

but

$$\frac{dy}{dx} = \lim_{(\Delta x \to 0)} \left[\frac{4 + \Delta x}{1} \right] = 4$$

Therefore,

$$\frac{dy}{dx} = 4 \quad \text{at } x = 2$$

This can be verified by examination of Fig. 4.

A general formula can be developed for evaluating dy/dx at any value of x by a method similar to the one used in the previous problem. If

$$y = x^2$$

then

$$y' = (x + \Delta x)^2$$

and

$$y' - y = \Delta y = (x + \Delta x)^2 - x^2$$

so that

$$\frac{\Delta y}{\Delta x} = \frac{(x + \Delta x)^2 - x^2}{\Delta x} = \frac{x^2 + 2x\Delta x + \Delta x^2 - x^2}{\Delta x}$$

$$= \frac{2x + \Delta x}{1} = 2x + \Delta x$$

Therefore

$$\frac{dy}{dx} = \lim_{(\Delta x \to 0)} \left[\frac{\Delta y}{\Delta x} \right] = 2x$$

or

$$\frac{d(x^2)}{dx} = 2x$$

which is a general expression that gives the value of dy/dx at any value of x. For example, at $x = +2$,

$$\frac{dy}{dx} = 2x = 4$$

which is the previous result.

The derivatives listed in Table I can be derived in a similar fashion. Each will be discussed.

TABLE I

SOME BASIC DERIVATIVES

$$\frac{dc}{dx} = 0 \quad (c \text{ is any constant})$$

$$\frac{d(x_n^n)}{dx} = nx^{n-1}$$

$$\frac{d(\sin x)}{dx} = \cos x$$

$$\frac{d(\cos x)}{dx} = -\sin x$$

$$\frac{de^x}{dx} = e^x$$

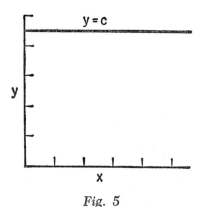

Fig. 5

Figure 5 shows the function $y = c$. Notice that even a large change in x at any point causes no change at all in y, so that $\Delta y/\Delta x = 0/\Delta x = 0$ at all points. Therefore, $dy/dx = 0$ also.

The expression $d(x^n)/dx$ is a general one which becomes $d(x^2)/dx$ if $n = 2$. It is suggested that the reader check this case to see that it agrees with the formula previously derived for $d(x^2)/dx$.

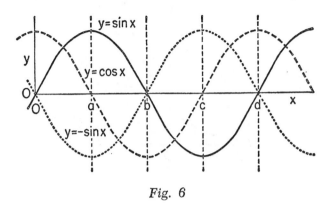

Fig. 6

Figure 6 illustrates $d(\sin x)/dx$ and $d(\cos x)/dx$. First examine $y = \sin x$, the solid curve. At points a and c it is flat (zero slope) so that at these points dy/dx must be zero. From a to c, $y = \sin x$ is sloped downward to the right with its steepest descent at $x = b$. Therefore, dy/dx is expected to be negative in this region, with its largest negative value at $x = b$, because the curve is steepest at that point. These conditions are met by the curve $y = \cos x$. Similarly $-\sin x$ can be shown to be the first derivative with respect to x of $y = \cos x$.

The function $y = e^x$ illustrated in Fig. 7 is a beauty. It is its own first derivative because it has a value equal to its slope at every point. This can be checked by drawing tangents at a few randomly selected points on the curve of Fig. 7. The slope of the curve will invariably equal y.

The fact that a function is continuous does not mean that it has a continuous first derivative. For example, Fig. 8 shows the first derivative of Fig. 3.

Notice that at $x = 1.5$ the first derivative is discontinuous. This means that the slope of Fig. 3 is different when approaching $x = 1.5$ from the left side than when approaching $x = 1.5$ from the right side.

The formulas given in Table I for evaluating various first derivatives are the basic group of a battery of several

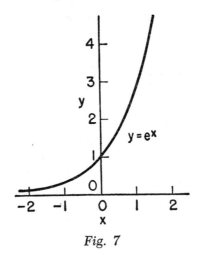

Fig. 7

dozen, found in any handbook, that permit the direct calculation of the derivative of any function that can be described with an equation.

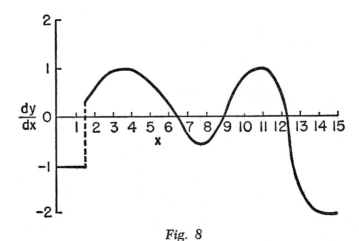

Fig. 8

DERIVATIVES OF INVOLVED EXPRESSIONS. Sometimes y is such an involved function of x that although each of the parts is simple to differentiate, their combination appears formidable. The use of a few simple rules simplifies the differentiation process. These rules are listed in Table II. In each case *a* is a constant, *u* is a function of *x* and *v* is another function of *x*.

TABLE II

DERIVATIVES OF INVOLVED EXPRESSIONS

Rule 1. $\dfrac{d(au)}{dx} = a\dfrac{du}{dx}$

Rule 2. $\dfrac{d(u + v)}{dx} = \dfrac{du}{dx} + \dfrac{dv}{dx}$

Rule 3. $\dfrac{d(u \cdot v)}{dx} = u\dfrac{dv}{dx} + v\dfrac{du}{dx}$

Rule 4. $\dfrac{d(u \div v)}{dx} = \dfrac{1}{v^2}\left(v\dfrac{du}{dx} - u\dfrac{dv}{dx}\right)$

$$a = \text{constant}$$
$$u = f_1(x)$$
$$v = f_2(x)$$

Rule 1

If $y = au$

$$\frac{dy}{dx} = a\frac{du}{dx}$$

This has been assumed already and used. For example,

$$y = 3x^2$$

here

$$u = x^2 \quad \text{and} \quad a = 3$$

and

$$\frac{dy}{dx} = \frac{3d(x^2)}{dx} = 3(2x) = 6x$$

Rule 2

If $y = u + v$

$$\frac{dy}{dx} = \frac{du}{dx} + \frac{dv}{dx}$$

EXAMPLE:

$$y = x^2 + e^x$$

here

$$u = x^2 \quad \text{and} \quad v = e^x$$

and

$$\frac{dy}{dx} = \frac{d(x^2)}{dx} + \frac{d(e^x)}{dx} = 2x + e^x$$

Rule 3

If $y = u \cdot v$

$$\frac{dy}{dx} = u\frac{dv}{dx} + v\frac{du}{dx}$$

EXAMPLE:

$$y = xe^x$$

here

$$u = x \quad \text{and} \quad v = e^x$$

and

$$\frac{dy}{dx} = \frac{xd(e^x)}{dx} + \frac{e^x d(x)}{dx} = xe^x + e^x$$

Rule 4

If $y = u/v$

$$\frac{dy}{dx} = \frac{1}{v^2}\left[v\frac{du}{dx} - u\frac{dv}{dx} \right]$$

EXAMPLE:

$$y = \frac{x^6}{x^5}$$

(In this case it is obvious that $y = x^{6-5} = x$ so that $dy/dx = 1$, but as an example, Rule 4 will be used to check it the hard way.)

Let

$$u = x^6 \quad \text{and} \quad v = x^5$$

Then

$$\frac{dy}{dx} = \frac{1}{(x^5)^2}\left[x^5\frac{d(x^6)}{dx} - x^6\frac{d(x^5)}{dx} \right]$$

$$= \frac{1}{x^{10}}\left[x^5(6x^5) - x^6(5x^4) \right]$$

$$= \frac{6x^{10} - 5x^{10}}{x^{10}} = \frac{x^{10}}{x^{10}} = 1$$

TANGENTS AND NORMALS. Often the equation of the tangent to a curve at a point is desired. The first derivative is of help in such problems.

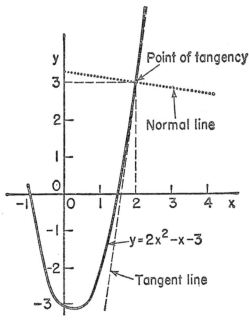

Fig. 9

For example, suppose that y is the function of x shown in Fig. 9 which can be expressed as

$$y = 2x^2 - x - 3 \qquad (1)$$

and the equation of the tangent line to the curve at $x = 2$ is desired. The equation of a line through any point (x_1, y_1) with a slope of m is

$$y - y_1 = m(x - x_1) \tag{2}$$

In this case, $x_1 = 2$ and y_1 is found by substituting the value $x = 2$ in Equation 1 to give:

$$y_1 = 2(2)^2 - 2 - 3 = 3$$

The slope of the tangent is the slope of the given curve and can be evaluated by finding the first derivative,

$$\frac{dy}{dx} = \frac{d(2x^2 - x - 3)}{dx} = 2(2x) - 1 + 0 = 4x - 1$$

and evaluating dy/dx at $x = 2$ to give:

$$\frac{dy}{dx} = 4(2) - 1 = 7 = m \quad (\text{at } x = 2)$$

so that m in the tangent line equation has the value 7.

The equation of the tangent line to the curve at the point $(2, 3)$ is now determined by inserting known values into Equation 2 to give:

$$y - 3 = 7(x - 2)$$

or

$$y = 7x - 11 \quad (\text{tangent line at } x = 2, y = 3)$$

which is drawn dashed in Fig. 9.

The normal line to a curve at a point is at right angles to the tangent line at the point and so has a slope that is the *negative reciprocal of the slope of the tangent*. The slope of the normal curve is therefore $-1/m$ and its equation is:

$$y - y_1 = \frac{-1}{m}(x - x_1) \tag{3}$$

In this example the equation of the normal curve at $(2, 3)$ is therefore

$$y - 3 = \frac{-1}{m}(x - 2) = \frac{-1}{7}(x - 2)$$

or

$$y = -0.143x + 3.286 \quad \text{(normal line at } x = 2, y = 3)$$

This is drawn dotted in Fig. 9.

MAXIMUM AND MINIMUM POINTS. Finding the point at which a curve has a maximum or minimum value is a common problem that calculus makes easy.

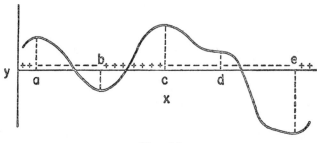

Fig. 10

In Fig. 10, points *a, b, c, d, e* are all points of zero slope. For a point to be a maximum or minimum point the first derivative (slope) must be zero at that point. But a further qualification is needed because points *a, b, c, e* are maximum or minimum points but *d* is not. Also, it would be nice to be able to tell directly from the equation of the curve that *a* and *c* are maximum points and that *b* and *e* are minimum points.

Indicated on the *x* axis of Fig. 10 are plus and minus signs to point out at what values of *x* the first derivative is positive and negative. Notice that at maximum points, such as *a* and *c*, the slope is positive to the left of the point and negative to the right of it. At minimum points, such as *b* and *e*, the slope is negative to the left of the point and positive to the right of it. At the inflection point *d* there is no change in the sign of the slope. This criterion could be used to classify points of zero slope as maximum, minimum or inflection points but it is a bit prosaic. It suggests that the classification of maximum and minimum depends on a change in first derivative. Just as the first derivative is a measure of the change in a function, perhaps a second derivative will be a measure of change in the first derivative that will lead to a means of directly separating maximum, minimum and inflection points.

Higher derivatives will now be introduced as a prelude to developing an analytic method of calculating maximum and minimum points.

HIGHER DERIVATIVES. The second derivative is defined as the first derivative of the first derivative:

$$\frac{d^2y}{dx^2} = \frac{d(dy/dx)}{dx}$$

and the third derivative is defined as the first derivative of the second derivative:

$$\frac{d^3y}{dx^3} = \frac{d(d^2y/dx^2)}{dx}$$

and so forth. A few examples may help develop a feeling for this process.

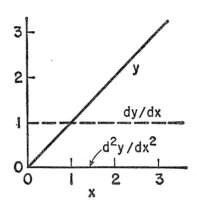

Fig. 11

Case I: (Fig. 11)

$$y = x$$

First derivative:

$$\frac{dy}{dx} = \frac{d(x)}{dx} = 1$$

Second derivative:

$$\frac{d^2y}{dx^2} = \frac{d(dy/dx)}{dx} = \frac{d(1)}{dx} = 0$$

Third derivative:

$$\frac{d^3y}{dx^3} = \frac{d(d^2y/dx^2)}{dx} = \frac{d(0)}{dx} = 0$$

In Fig. 11 the given function, y, is plotted as a solid line. As its slope is constant at a value of 1, the first derivative (dashed curve) is constant for all values of x. As the second derivative is the first derivative of the first derivative (or slope), and the first derivative is a line of zero slope, the second derivative is zero. Any higher derivatives are zero also.

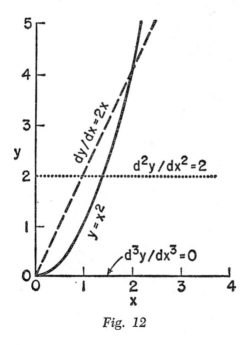

Fig. 12

Case II: (Fig. 12)

$$y = x^2$$

First derivative:

$$\frac{dy}{dx} = \frac{d(x^2)}{dx} = 2x$$

Second derivative:

$$\frac{d^2y}{dx^2} = \frac{d(dy/dx)}{dx} = \frac{d(2x)}{dx} = 2$$

Third derivative:

$$\frac{d^3y}{dx^3} = \frac{d(d^2y/dx^2)}{dx} = \frac{d(2)}{dx} = 0$$

In Fig. 12 the function, $y = x^2$ (solid curve), has a slope that is increasingly steep for increasing values of x. It is not surprising therefore to see that its first derivative (dashed line) increases as x increases. The first derivative is a straight line (constant slope), so the second derivative is a constant and the third derivative is zero.

Case III: (Fig. 13)

$$y = x^3 \tag{4}$$

First derivative:

$$\frac{dy}{dx} = \frac{d(x^3)}{dx} = 3x^2$$

Second derivative:

$$\frac{d^2y}{dx^2} = \frac{d(dy/dx)}{dx} = \frac{d(3x^2)}{dx} = 6x \tag{5}$$

Third derivative:

$$\frac{d^3y}{dx^3} = \frac{d(d^2y/dx^2)}{dx} = \frac{d(6x)}{dx} = 6$$

Fourth derivative:

$$\frac{d^4y}{dx^4} = \frac{d(d^3y/dx^3)}{dx} = \frac{d(6)}{dx} = 0$$

This last example of functions of the form $y = x^n$ can be compared to $y = x^2$. Except for multiplication by a factor of

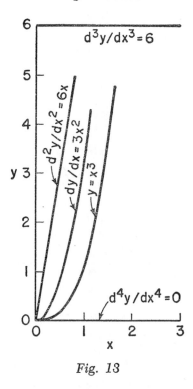

Fig. 13

3, x^2 is the first derivative of x^3 and so the second derivative of x^3 corresponds to the first derivative of x^2, etc.

As an example of the evaluation of higher derivatives at particular values of the independent variable, the second derivative of $y = x^3$ when $x = 3$ is found by substituting the particular value of x (3 in this case) into the general expression of d^2y/dx^2 which was stated in Equation 5 to give

$$\left[\frac{d^2y}{dx^2}\right]_{(x=3)} = 6(3) = 18$$

(Notice the bracket notation used to express "the value of d^2y/dx^2 when $x = 3$.")

If the problem above were to evaluate d^2y/dx^2 when $y = 12$, it would be necessary to first find the value of x that corresponds to $y = 12$ by substituting 12 for y in Equation 4 to find that $x = 2.29$ when $y = 12$. The value of the second derivative of y with respect to x at $y = 12$ is therefore:

$$\left[\frac{d^2y}{dx^2}\right]_{(y=12)} = 6(2.29) = 13.74$$

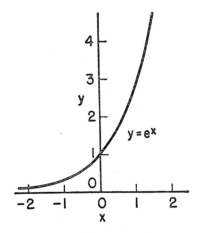

Fig. 7 (*repeated*)

Case IV: (Fig. 7)

$$y = e^x$$

First derivative:

$$\frac{dy}{dx} = \frac{d(e^x)}{dx} = e^x$$

Second derivative:

$$\frac{d^2y}{dx^2} = \frac{d(e^x)}{dx} = e^x$$

In fact, all succeeding derivatives of e^x are e^x. As previously stated, e^x is its own derivative. It is the only function except for zero that has that distinction.

Case V:

$$y = \sin x$$

It is suggested that the reader try to find the fourth derivative of $y = \sin x$ as an exercise. The answer is $d^4y/dx^4 = \sin x$. The answer can be derived by using the formulas of Table I and differentiating four successive times.

CALCULATION OF MAXIMUM AND MINIMUM POINTS. At any stated value of x, the first derivative of a function of x describes by its sign and number value the slope of the function at that point. In a similar manner, the second derivative defines curvature and can be used to separate maximum, minimum and inflection points.

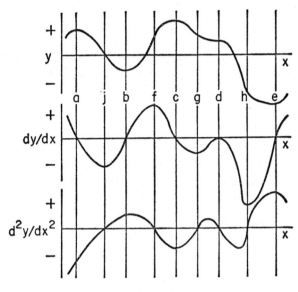

Fig. 14

Figure 14 shows the function of Fig. 10 redrawn with its first and second derivatives drawn below. These additional curves demonstrate how maximum and minimum points can be determined from first and second derivatives:

(a) At a maximum point the first derivative is zero and the second derivative is negative. (Notice points a and c of Fig. 14.)

(b) At a minimum point the first derivative is zero and the second derivative is positive. (Note points b and e of Fig. 14.)

As a numerical example, the maximum and minimum points of the following expression will be calculated:

$$y = x^3 + 4x^2 - x - 4 \qquad (6)$$

Possible maximum and minimum points are found by first calculating dy/dx,

$$\frac{dy}{dx} = 3x^2 + 8x - 1 \qquad (7)$$

and then finding values of x for which $dy/dx = 0$, i.e., those values of x for which:

$$0 = 3x^2 + 8x - 1 \qquad (8)$$

By trial and error, by graphing dy/dx and inspecting the graph, or by using the quadratic formula,* the following values of x can be found to satisfy Equation 8:

$$x = +0.120 \qquad (9)$$
$$x = -2.786$$

These values of x locate possible maximum or minimum points of Equation 6. Computing the second derivative of Equation 6 by computing the first derivative of Equation 7 gives

$$\frac{d^2y}{dx^2} = \frac{d(dy/dx)}{dx} = \frac{d(3x^2 + 8x - 1)}{dx}$$

or

$$\frac{d^2y}{dx^2} = 6x + 8 \qquad (10)$$

Evaluating Equation 10 at the possible minimum and maximum locations produces:

$$\left[\frac{d^2y}{dx^2}\right]_{(x=+.120)} = 6(.120) + 8 = 8.72 \qquad (11)$$

$$\left[\frac{d^2y}{dx^2}\right]_{(x=-2.786)} = 6(-2.786) + 8 = -8.716$$

Combining the data from Equations 9 and 11, it can be concluded that there is a maximum of Equation 6 at $x = -2.786$

---

* The quadratic formula states: If $ax^2 + bx + c = 0$, then
$$x = \frac{-b \pm \sqrt{b^2 - 4ac}}{2a}.$$

since at that value the first derivative of y is zero and the second derivative of y is negative; there is a minimum of Equation 6 at $x = +.120$ because at that value the first

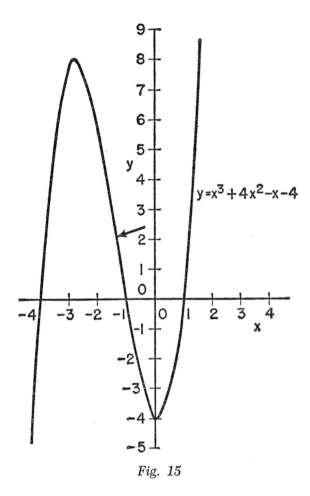

$y = x^3 + 4x^2 - x - 4$

Fig. 15

derivative of y is zero and the second derivative of y is positive. These values are correct as can be seen by examining the graph of Equation 6 as shown in Fig. 15.

CURVATURE AND INFLECTION. As the first derivative indicated slope, the second derivative indicates curvature.

Notice in Fig. 14 that wherever the function has a positive second derivative it is concave upward and wherever it has a negative second derivative it is concave downward. For example, from j to f, g to d, and h to e the second derivative is positive and the y curve is concave upward. From f to g and from d to h the second derivative is negative and the y curve is concave downward.

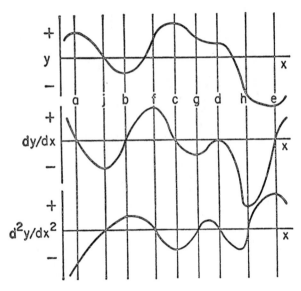

Fig. 14 (*repeated*)

Points where the direction of curvature changes are called inflection points. Here the second derivative is zero. Examples in Fig. 14 are at j, f, g, d, and h.

As an example of the use of calculus to find an inflection point, the location of the inflection point of Equation 6 will be computed. Equation 6 was stated as:

$$y = x^3 + 4x^2 - x - 4 \tag{6}$$

and its second derivative was calculated to be

$$\frac{d^2y}{dx^2} = 6x + 8 \tag{10}$$

Setting Equation 10 equal to zero gives:

$$6x + 8 = 0$$

or

$$x = -1.333 \quad \text{(location of inflection point)}$$

This is correct and can be checked by examining Fig. 15 where the inflection point is indicated by a short arrow. Notice that the inflection point marks the transition between a part of the graph that is concave upward and a part of the graph that is concave downward.

Of interest at times is an accurate measure of the radius of curvature of a function at a point. If y is a function of x, then the radius of curvature, r, at any point is given by

$$r = \frac{[1 + (dy/dx)^2]^{3/2}}{d^2y/dx^2} \tag{12}$$

Notice that both the first and second derivatives enter into this formula unless the slope at that point is zero. For this particular case Equation 12 becomes:

$$r = \frac{1}{d^2y/dx^2} \tag{13}$$

As examples of radius of curvature calculations, the radius of curvature of Equation 6 (Fig. 15) will be calculated for $x = -1$, $x = -1.33$ (the inflection point), and $x = -2.786$ (the maximum point). At $x = -1$, the first and second derivatives of Equation 6 are found by putting number values into Equations 7 and 10 to give:

$$\frac{dy}{dx} = 3x^2 + 8x - 1 \tag{7}$$

or

$$\frac{dy}{dx} = +3 - 8 - 1 = -6 \quad \text{(at } x = -1)$$

and

$$\frac{d^2y}{dx^2} = 6x + 8 \tag{10}$$

or

$$\frac{d^2y}{dx^2} = -6 + 8 = +2 \quad \text{(at } x = -1)$$

Inserting these values into Equation 12 yields:

$$r = \frac{(1 + 6^2)^{3/2}}{2} = \frac{(37)^{3/2}}{2} = \frac{224}{2} = 112 \text{ units}$$

which means that in the immediate vicinity of $x = -1$ the curve could be fit by the arc of a circle 62.5 units in radius.

If the radius calculation is repeated for the case of the maximum point at $x = -2.786$, since $dy/dx = 0$ at this point and $d^2y/dx^2 = -8.716$ as calculated previously, Equation 12 yields:

$$r = \frac{(1 + 0)^{3/2}}{-8.716} = \frac{1}{-8.716} = .1147$$

A very small radius indeed! Note also that since dy/dx was zero in the last example, the calculation degenerated into that of Equation 13.

Since d^2y/dx^2 is zero at an inflection point, the radius of curvature there is

$$r = \frac{(1 + (dy/dx)^2)^{3/2}}{0} = \infty$$

As an exercise the reader is invited to calculate the maximums, minimums and inflection points of $y = -\sin x$ for the range of values of x from zero to six radians.* The results can be checked by examination of Fig. 6.

DIMENSIONAL UNITS. In problems involving physical quantities the dependent and independent variables have dimensional units. Similarly the derivatives have dimensional units which follow the rule:

$$\text{Dimensions of } \frac{d^ny}{dx^n} = \frac{\text{dimensions of } y}{(\text{dimensions of } x)^n}$$

For example, if y is in [pounds] and x is in [inches], dy/dx has units of [pounds/inch] and d^3y/dx^3 has units of [pounds per cubic inch].

* One radian = 57.3 degrees.

2. ELEMENTARY DIFFERENTIAL EQUATIONS

INTERPRETING DIFFERENTIAL EQUATIONS. Differential equations often arise when physical problems are put into mathematical form. As a very crude example, if it is desired to write the equations that describe the motion of a steel ball in an empty soup bowl, the force that accelerates the ball toward the center of the bowl is a function of the slope of the bowl wall. The slope of the bowl wall is, of course, the first derivative of the curve that describes the shape of the bowl. Thus in this problem the equations of motion will include a first derivative term. Other examples will be given in more detail in Chapter 5.

"Solving" an equation, like solving a jigsaw puzzle, means rearranging the parts to form a more readily understood relationship than exists in the problem or puzzle in its original organization. A differential equation is usually considered to be solved when the relationship between dependent and independent variables is in the form of a graph or a simple analytic expression containing no derivatives or integrals.* Solving differential equations is both a science and an art and encompasses a field that even includes analog and digital computers. Since this chapter is primarily designed to introduce the concepts of differential equations, most of the discussion of methods used to solve differential equations is postponed to Chapter 6. The discussion of the solution of a few differential equations is presented here to help develop a fundamental understanding of differential equations.

* "Integral" will be defined in Chapter 3.

Case I:

$$\frac{dy}{dx} = 3$$

Since dy/dx is a definition for "slope," the given differential equation simply states that the curve of y vs x has a constant slope of 3. This means that a plot of y as a function of x will be a straight line whose slope at all points is 3.

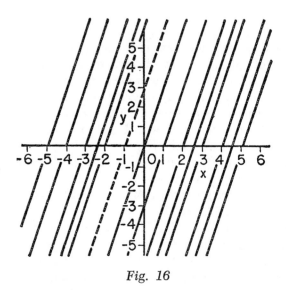

Fig. 16

This is true of all the lines shown in Fig. 16; therefore, any one of them is a solution of the given equation. If an additional condition were stated, such as $y = 3$ at $x = 0$, only one of the family of solutions would satisfy both the equation and the additional condition. This particular case is the dashed line of Fig. 16. "Additional conditions" are also called "auxiliary conditions," "initial conditions" and "boundary conditions"; the latter two are the more common terms.

Case II:

$$\frac{dy}{dx} + 2x = 0 \tag{14}$$

boundary conditions:

$$y = 1 \text{ at } x = 0.$$

In this case one point on the curve has been given. By inserting the value $x = 0$ in the equation one can discover that dy/dx (the slope) is zero at $x = 0$.

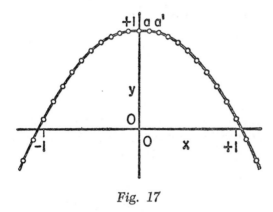

Fig. 17

This information is summarized in line segment a-a' in Fig. 17. At a', $x = 0.1$ and so by substitution in Equation 14

$$\frac{dy}{dx} + 2(0.1) = 0 \quad (\text{at } a')$$

or

$$\frac{dy}{dx} = -0.2 \qquad (\text{at } a').$$

The segment of the curve to the right of a' is drawn at a slope of -0.2 until $x = 0.2$ where the slope is again recalculated and a new curve segment is drawn. By this means the entire curve can be drawn, giving the desired curve of y as a function of x. This method is only an approximate one, for dy/dx varies continuously with x, not in discrete steps of 0.1 unit changes in x as shown. If smaller steps are taken, a better approximation is obtained. A judgment of the approximation made in this case is possible by comparing Fig. 17 with Fig. 18 which is an exact solution.

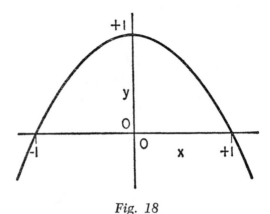

Fig. 18

As the above example indicates, a differential equation
and its boundary conditions merely express a curve in terms
of its point-by-point characteristics.

Case III:

$$\frac{d^2y}{dx^2} - 2\frac{dy}{dx} = 0 \tag{15}$$

Equation 15 is called a second-order differential equation
because the second derivative is the highest order derivative appearing in the equation. Similarly, the previous case,
Case II, was a first-order differential equation. In general,
every differential equation must have the same number of
boundary conditions as its order number if it is to have a
single curve as a solution. In this case it is assumed that the
boundary conditions were given as $y = 0$ and $d^2y/dx^2 = 2$
at $x = 0$. From the boundary conditions the curve passes
through the point $(0, 0)$. This is point a in Fig. 19. Inserting the initial condition, $d^2y/dx^2 = 2$, into Equation 15
results in:

$$2 - 2\frac{dy}{dx} = 0 \qquad \text{(at } x = 0\text{)}$$

or

$$\frac{dy}{dx} = +1 \quad \text{(at } x = 0\text{)}$$

Therefore, line segment *ab* is drawn at a slope of $+1$. As

$$\frac{d^2y}{dx^2} = 2 \quad (\text{at } x = 0)$$

then also

$$\frac{d(dy/dx)}{dx} = 2 \quad (\text{at } x = 0)$$

so that for an increase of 0.1 in *x*, the slope of the first derivative will increase 2×0.1 or $+0.2$. One can therefore assume a slope of 1.2 for the region just before point *c*.

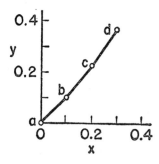

Fig. 19

Equation 15 can be rewritten as

$$\frac{d^2y}{dx^2} = \frac{2dy}{dx}$$

or

$$\frac{d(dy/dx)}{dx} = \frac{2dy}{dx}$$

and as an approximation this can be converted to

$$\frac{\Delta(dy/dx)}{\Delta x} = 2dy/dx$$

or

$$\Delta(dy/dx) = 2(dy/dx)\Delta x$$

which states that the change in slope of the curve during a

small change in x is equal to twice the slope times the change in x, so that during the 0.1 increase in x from point c to point d, the slope is its prior value of 1.2 plus an increase of 2.4 × 0.1, or 1.2 + 0.24 = 1.44. Similarly the entire curve could be developed. As before, the smaller the x increments used, the more accurate the curve that results.

The purpose of presenting the above methods of solving differential equations was not to indicate the best way or necessarily a typical way they are solved, but to demonstrate that differential equations have simple specific meanings.

II

INTEGRAL CALCULUS

3. SIMPLE INTEGRATION

SUMMATION. Figure 20 shows a scale that measures a quantity x. In Fig. 20 y is a length cut out of a piece of rope marked off in units of x. The dashed lines indicate where section y was removed from the original rope. The length of y can be found by cutting it into pieces of length Δx and adding these pieces together as indicated in Fig. 20A.

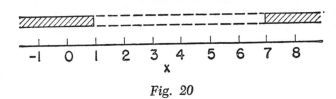

Fig. 20

Stated in symbol form:

$$y = \sum_{x=1}^{x=7} \Delta x$$

Here the Greek letter "sigma" stands for the word "sum" and the notation means: $y =$ the sum of the lengths of all the Δx pieces of rope which are laid end to end from $x = 1$ to $x = 7$. The total length, y, is the same whether each Δx

is a long segment (Fig. 20A) or a small segment (Fig. 20B). Either case includes all the rope segments between the "limits" of summation of $x = 1$ and $x = 7$.

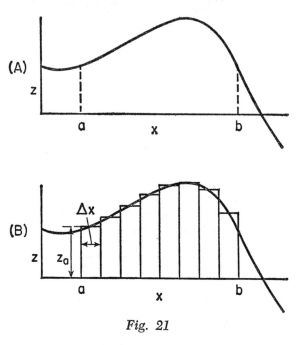

Fig. 21

A second example of summation, Fig. 21A, shows a curve of z as a function of x. As before, x can be divided into little segments, Δx, as shown in Fig. 21B. Examine the expression

$$y = \sum_{x=a}^{x=b} z \, \Delta x \tag{16}$$

Again this is a sum, but instead of being a sum of Δx lengths, it is the sum of the product of each Δx multiplied by the height of the curve z. It is implied in the notation that for each Δx the value of z to be used is the one corresponding to the location on the x axis of the midpoint of that particular Δx. For example, the first Δx to the right of $x = a$ is multiplied by z_a, the number value of z at the center of the corresponding Δx strip. Each $z \, \Delta x$ product can be drawn as a rectangle and y becomes the sum of the individual rectangular areas shown in Fig. 21B so that y approximates the area under curve z between end points $x = a$ and $x = b$.

In conclusion then, if z is a function of x and

$$y = \sum_{x=a}^{x=b} z\,\Delta x \tag{16}$$

then $y =$ the approximate area under the given curve of z vs x as bounded by the x axis and the lines $x = a$ and $x = b$. As will be demonstrated later the word "approximate" can be removed by taking very small values of Δx.

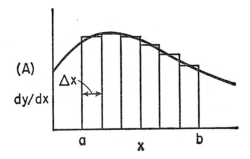

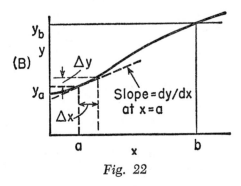

Fig. 22

Figure 22 shows a third example of summation where z is a very particular function, dy/dx. Again the summing operation

$$\sum_{x=a}^{x=b} \left(\frac{dy}{dx}\right) \Delta x$$

can be depicted, as shown in Fig. 22A, as the sum of the areas of all the rectangles shown from $x = a$ to $x = b$ where each rectangle's height is the value of dy/dx at its midpoint and each rectangle's width is Δx. It will be demonstrated by Fig. 22B that this summation can be interpreted as the curve of y vs x from which the dy/dx curve was derived originally.

By an earlier definition dy/dx is the limiting value of the slope $\Delta y/\Delta x$ as Δx approaches zero. If

$$dy/dx = \text{approximately } \Delta y/\Delta x$$

then

$$(dy/dx)\,\Delta x = \text{approximately } (\Delta y/\Delta x)\,\Delta x$$

$$= \text{approximately } \Delta y$$

which is the change in y corresponding to a change in x of Δx. The summation of the rectangles in Fig. 22A is in reality the sum of all the changes in y as x goes from a to b, or

$$\sum_{x=a}^{x=b} \left(\frac{dy}{dx}\right)\Delta x = \sum_{x=a}^{x=b} \Delta y = y_{(\text{at } x=b)} - y_{(\text{at } x=a)} = y_b - y_a$$

In other words the summation process is the inverse of differentiation. Just as differentiating y yields dy/dx, a summation process of dy/dx will approximate y. But before the summation idea can be used to solve problems it must be refined from an approximate process to an exact process called *integration*.

INTEGRATION. In the basic summation process

$$y = \sum_{x=a}^{x=b} z\,\Delta x \qquad (16)$$

as Δx is taken smaller and smaller to approach the limit of zero, y becomes the exact area under the z curve between $x = a$ and $x = b$. For this condition $(\Delta x \rightarrow 0)$, the symbol dx replaces the symbol Δx. The summation process now becomes a continuous process between $x = a$ and $x = b$ so the discontinuous symbol "sigma" is replaced by a smooth flowing tilted letter $\int$, thus transforming the basic summation formula, Equation 16, into the following basic equation for integration:

$$y = \int_{=a}^{x=b} z \, dx$$

and now y is the exact area under the curve of z vs x from $x = a$ to $x = b$. Similarly the summation expression

$$y_b - y_a = \sum_{x=a}^{x=b} \left(\frac{dy}{dx}\right) \Delta x \quad \text{(approximately)}$$

becomes the integral expression

$$y_b - y_a = \int_{x=a}^{x=b} \left(\frac{dy}{dx}\right) dx \quad \text{(exactly)} \qquad (17)$$

Two examples will be used to demonstrate the application of Equation 17.

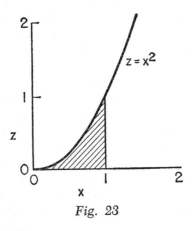

Fig. 23

EXAMPLE I: (Fig. 23)—Find the area (shaded) under the curve $z = x^2$ between $x = 0$ and $x = 1$.

$$\text{Area} = \int_{x=0}^{x=1} z \, dx = \int_{x=0}^{x=1} x^2 \, dx \qquad (18)$$

This problem can be solved if a function, y, can be found such that its first derivative, dy/dx, equals x^2. From a previous problem, if

$$y = \frac{1}{3}(x^3) = \frac{x^3}{3}$$

then

$$\frac{dy}{dx} = \frac{1}{3}(3x^2) = x^2$$

Substituting the above values for y and z:

$$\text{Area} = \int_{x=0}^{x=1} x^2\, dx = \int_{x=0}^{x=1} \left(\frac{dy}{dx}\right) dx \qquad (19)$$

$$\text{Area} = y_{(x=1)} - y_{(x=0)} \qquad (20)$$

$$\text{Area} = \left[\frac{x^3}{3}\right]_{(x=1)} - \left[\frac{x^3}{3}\right]_{(x=0)} = \frac{1}{3} - 0 = \frac{1}{3} \qquad (21)$$

This typical problem shows the steps essential to any integration. *Except for a direct graphical measurement of the area under a curve to evaluate an integral, all integration requires that the function being integrated must be recognized as the derivative of another function.* In Equation 18, the function, z, was first translated into a function of x. x is called the variable of integration. Notice that the limits, $x = 1$ and $x = 0$, are also expressed in terms of the variable of integration. Equations 19 and 20 show that the procedure for evaluating the integral depends upon finding (from memory or a handbook) some function of x whose derivative happens to be the function one is trying to integrate. Once this function is found, it is evaluated at the upper limit and lower limit (Equation 21) and the value of the integral is the difference between the upper and lower limit values. Equation 21 is sometimes shown as

$$\text{Area} = \left[\frac{x^3}{3}\right]_{x=0}^{x=1} \quad \text{or} \quad \text{Area} = \left[\frac{x^3}{3}\right]_{0}^{1} = \frac{(1)^3}{3} - \frac{(0)^3}{3} = \frac{1}{3}$$

EXAMPLE II: (Fig. 24)—Find the area shaded under the curve,

$$z = \cos x$$

between the limits $x = \pi/2$ and $x = 3\pi/2$. (NOTE: x is in radians, a requirement for all calculus operations with trigonometric functions: 2π radians = 360 degrees; $\pi/2$ radians = 90 degrees; $3\pi/2$ radians = 270 degrees.)

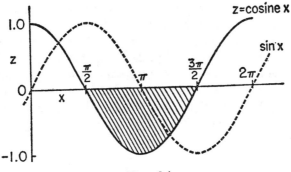

Fig. 24

Proceeding as before,

$$\text{Area} = \int_{x=\pi/2}^{x=3\pi/} z \, dx = \int_{x=\pi/2}^{x=3\pi/2} \cos x \, dx$$

but

$$\cos x = \frac{d(\sin x)}{dx}$$

So in this case:

$$\text{Area} = \left[\sin x \right]_{x=\pi/2}^{x=3\pi/2} = \sin \frac{3\pi}{2} - \sin \frac{\pi}{2}$$
$$= \sin 270° - \sin 90°$$

Sin x has been shown dashed in Fig. 24 to aid in evaluating the function at its limits giving

$$\text{Area} = -1 - (+1) = -2$$

The area is negative because it lies below the x axis. The integral of a sine or cosine function may include areas both above (positive) and below (negative) the axis. At times during integration the positive and negative areas will cancel each other. As an example the reader should find the area under the curve $z = \cos x$ between the limits $x = 0$ and $x = 2\pi$. Proceeding as in the previous example should cause the net area to equal zero.

LIMITS AND CONSTANTS OF INTEGRATION. The integrals discussed so far have been *definite* integrals in which integration has been performed between definite limits. Often equations are stated in *indefinite* form, that is without limits —but with a *constant of integration* instead. As will be shown, this leads to a more simplified form of mathematical notation—but one that can be misleading if not properly interpreted. To illustrate this point, the basic equation of Fig. 24,

$$\text{Area} = \int_{x=a}^{x=b} \cos x \, dx$$

will be evaluated for several cases with the same lower limit but with different upper limits.

(1) $\text{Area} = \displaystyle\int_{x=\pi/4}^{x=\pi/2} \cos x \, dx = [\sin x]_{x=\pi/4}^{x=\pi/2} = 1 - 0.707 = 0.293$

(2) $\text{Area} = \displaystyle\int_{x=\pi/4}^{x=3\pi/4} \cos x \, dx = [\sin x]_{x=\pi/4}^{x=3\pi/4} = 0.707 - 0.707 = 0$

(3) $\text{Area} = \displaystyle\int_{x=\pi/4}^{x=\pi} \cos x \, dx = [\sin x]_{x=\pi/4}^{x=\pi} = 0 - 0.707 = -0.707$

The function and the areas calculated are shown in Fig. 25A. Figure 25B shows the general solution which includes the three specific solutions between the limits of $x = \pi/4$ and $x = \pi/2$, $3\pi/4$, and π. Figure 25B would have the same shape for any lower limit (assuming it were the same for all integrations), however a change in lower limit would cause a vertical shift of the curve. Figure 25C shows the same integral but with the lower limit set at a fixed value, k, and with a constant, c, added to the integral. In this case c has been chosen as the value of the integrated function at its lower limit. For example, if $k = \pi/4$, then

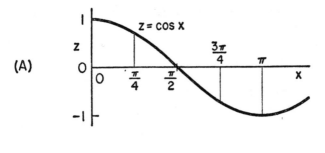

(A)

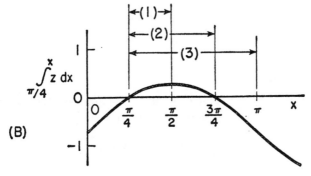

(B)

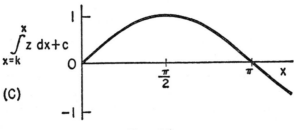

(C)

Fig. 25

$c = sin \, \pi/4$ or 0.707.* Constant c is called the *constant of integration* and by its use the integral of a function is itself a single-valued function.

* As long as $c = \sin k$ the curve of Fig. 24C will define the integral for any value of k. If $k = 0$, $c = 0$ or if $k = \pi/2$, $c = 1$; etc.

A look at the basic integration formula will explain this further.

$$y_b - y_a = \int_{x=a}^{x=b} \left(\frac{dy}{dx}\right) dx \qquad (17)$$

can be rewritten as:

$$y_b = \int_{x=a}^{x=b} \left(\frac{dy}{dx}\right) dx + y_a$$

or

$$y_b = \int_{x=a}^{x=b} \left(\frac{dy}{dx}\right) dx + c \qquad (22)$$

where

$$c = y_a$$

By using the constant c, y_b is a function only of the upper limit $x = b$, and Equation 22 can be simplified to:

$$y_b = \int^{x=b} \left(\frac{dy}{dx}\right) dx + c \qquad (23)$$

But the notation y_b and $x = b$ can be assumed, so Equation 23 can be further simplified to:

$$y = \int \left(\frac{dy}{dx}\right) dx + c \qquad (24)$$

This shorthand form is in such common use that it is often assumed, incorrectly, to be the basic definition of integration. Philosophically, Equation 24 points out that integration is the inverse of differentiation. Furthermore, if the value of c is known, the integral, y, is a single-valued function of the upper limit x.

Equation 24 can be used, for example, to find y when

$$\frac{d^2y}{dx^2} = 3x \qquad (25)$$

Subscripts will be used to discern between different constants of integration. Since

$$\int \left(\frac{dy}{dx}\right) dx + c_1 = y \qquad (26)$$

then by logical extension

$$\int \left(\frac{d^2y}{dx^2}\right) dx + c_2 = \frac{dy}{dx} \tag{27}$$

By substitution of Equation 25 into Equation 27

$$\frac{dy}{dx} = \int (3x)\, dx + c_2 = \frac{3x^2}{2} + c_2$$

because

$$\frac{d(3x^2/2)}{dx} = \frac{3}{2}(2x) = 3x$$

Therefore

$$y = \int \left(\frac{3x^2}{2} + c_2\right) dx + c_1 \tag{28}$$

From differential calculus

$$\frac{d(x^3/2)}{dx} = \frac{3x^2}{2} \tag{29}$$

and

$$\frac{d(c_2x)}{dx} = c_2 \tag{30}$$

so combining Equations 29 and 30 gives

$$\frac{d(x^3/2 + c_2x)}{dx} = \left(\frac{3x^2}{2} + c_2\right) \tag{31}$$

Thus by using the right side of Equation 31 as dy/dx in Equation 26, the answer to the problem is

$$y = x^3/2 + c_2x + c_1 \tag{32}$$

Notice that c_1 and c_2 may be any constant numbers. In any physical problem c_1, c_2 and, in fact, all constants of integration are defined as particular values by problem initial conditions or boundary values.

The solution (Equation 32) can be checked by finding the second derivative of y. If

$$y = x^3/2 + c_2x + c_1 \tag{32}$$

then

$$\frac{dy}{dx} = 3x^2/2 + c_2 + 0$$

and

$$\frac{d^2y}{dx^2} = 3x + 0 + 0$$

which checks Equation 25.

In summary, while the basic integration formula is often expressed as

$$y = \int \left(\frac{dy}{dx}\right) dx + c \qquad (24)$$

it can be interpreted in two ways:

The first is

$$y_b = \int_{x=a}^{x=b} \left(\frac{dy}{dx}\right) dx + y_a$$

or

$$y_b - y_a = \int_{x=a}^{x=b} \left(\frac{dy}{dx}\right) dx \qquad (17)$$

where $y_b - y_a =$ the area under the curve of dy/dx vs x. The second way is

$$y_b = \int_{x=k}^{x=b} \left(\frac{dy}{dx}\right) dx + c$$

where y and dy/dx are single-valued functions of x and c is a constant related to k.

TABLES OF INTEGRALS. Since integration depends on recognizing that a given function is a derivative of another function, an indirect process, tables of integrals are available, some of which are good sized books. A few of the more commonly used integrals are shown in Table III.

An example will illustrate the use of the formulas in Table III. If a force is applied to a spring to stretch it and the spring stretches ⅜ centimeters per dyne of force applied, what is the total energy in dyne-centimeters that is stored

TABLE III

COMMONLY USED INTEGRALS

$$\int x^n \, dx = \frac{x^{n+1}}{n+1} + c \quad (n \neq -1)$$

$$\int \frac{dx}{x} = \log_e x + c$$

$$\int a^x \, dx = \frac{a^x}{\log_e a} + c$$

$$\int e^x \, dx = e^x + c$$

$$\int \sin x \, dx = -\cos x + c$$

$$\int \cos x \, dx = \sin x + c$$

$$\int \sin^2 x \, dx = \tfrac{1}{2}(x - \sin x \cos x) + c$$

$$\int \cos^2 x \, dx = \tfrac{1}{2}(x + \sin x \cos x) + c$$

$$\int e^{ax} \, dx = \frac{e^{ax}}{a} + c$$

x is a variable; n is an integer; c and a are numbers.

in the spring if it is stretched 5 cm? 10 cm? 15 cm? Since energy stored is force (F) times the distance (x) that the force is applied, the additional energy stored during any microscopic interval of stretching, dx, is $F \, dx$ and the total energy (E) stored is:

$$E = \int_{x=0}^{x=b} F \, dx + c \tag{33}$$

where c is the constant of integration and b is the distance that the spring is stretched. The force is related to the amount of stretching by $F = 3x$ so that Equation 33 becomes:

$$E = \int_{x=0}^{x=b} 3x \, dx + c$$

Using line one of Table III reduces this integral to

$$E = \left[\frac{3x^2}{2} \right]_{x=0}^{x=b} + c$$

which expands to

$$E = \frac{3b^2}{2} - 0 + c \qquad (34)$$

Since the stored energy is zero if the spring is unstretched, $E = 0$ when $b = 0$ so that c must equal zero. This reduces Equation 34 to

$$E = \frac{3b^2}{2} \qquad (35)$$

Inserting values of 5 cm, 10 cm and 15 cm for b in Equation 35 produces answers for E of 37.5 dyne-cm, 150 dyne-cm and 337.5 dyne-cm respectively.

DIMENSIONAL UNITS. In an integral relationship such as

$$y = \int z \, dx + c$$

the dimensional units of y are the product of the dimensional units of z times the dimensional units of x; c has the same dimensional units as y. In the problem example above (Equations 33 through 35), E has units of [dyne-centimeters], x and b have units of [centimeters] and the equations seem to be dimensionally inconstant unless one remembers that the number "3" has dimensions of [dynes/centimeter]!!!

LAWS OF INTEGRALS. Laws that govern the combining and factoring of integrals are similar to the laws that govern the addition and multiplication of derivatives. The basic rules are listed in Table IV.

TABLE IV

LAWS OF INTEGRALS

Rule 1. $\displaystyle\int az\,dx = a\int z\,dx$

Rule 2. $\displaystyle\int (z+y)dx = \int z\,dx + \int y\,dx$

Rule 3. $\displaystyle\int_{x=a}^{x=c} z\,dx = \int_{x=a}^{x=b} z\,dx + \int_{x=b}^{x=c} z\,dx$

Rule 4. $\displaystyle\int_{x=a}^{x=b} z\,dx = -\int_{x=b}^{x=a} z\,dx$

Symbols x, y, and z are variables. a is a number.

Rule 1:

$$\int az\,dx = a\int z\,dx$$

This rule states that, as in the case of differentiation, multiplication by a constant can be done either inside or outside the calculus operation without changing the value of the total expression.

For example,

$$2\pi\int \sin 3x\,dx = \int 2\pi \sin 3x\,dx$$

Rule 2:

$$\int (z+y)\,dx = \int z\,dx + \int y\,dx$$

As integration is the inverse of differentiation it is not surprising that the same rule for the sum of several terms applies. This rule can be used to readily integrate involved expressions that can be broken into partial fractions. For example,

$$y = \int \frac{(x+1)(x+3)}{x^3}\,dx$$

can be integrated by reducing the integral to:

$$y = \int \frac{x^2 + 4x + 3}{x^3}\, dx = \int \left(\frac{1}{x} + \frac{4}{x^2} + \frac{3}{x^3} \right) dx$$

and applying Rule 2

$$y = \int \frac{1}{x}\, dx + \int \frac{4}{x^2}\, dx + \int \frac{3}{x^3}\, dx$$

from which Table III (lines one and two) indicates that

$$y = \log_e x - \frac{4}{x} - \frac{3}{2x^2} + c$$

Notice that only one constant of integration is shown since the sum of three constants is a single constant.

Rule 3:

$$\int_{x=a}^{x=c} z\, dx = \int_{x=a}^{x=b} z\, dx + \int_{x=b}^{x=c} z\, dx \qquad (36)$$

This statement means, for example, that the integral (area under the curve) of z in Fig. 25A from $x = \pi/4$ to $x = 3\pi/4$ is the sum of the integral of z from $x = \pi/4$ to $x = \pi/2$ and the integral of z from $x = \pi/2$ to $x = 3\pi/4$. If integration is thought of as an area summing process this rule is almost intuitively obvious. It is often a great time saver when the curve has symmetry as in the example stated above, for then one integration of one of the two terms on the right-hand side of Equation 36 can be performed and the result doubled to evaluate the left side of Equation 36.

Rule 4:

$$\int_{x=a}^{x=b} z\, dx = - \int_{x=b}^{x=a} z\, dx$$

All integrations performed up to this point have been implied to be from left to right along the x axis, i.e., in the direction of increasing values of x. Rule 4 states that if the direction is reversed, the integral changes sign. This can be seen by examining the integral expression,

$$y_b - y_a = \int_{x=a}^{x=b} (dy/dx)\, dx \qquad (17)$$

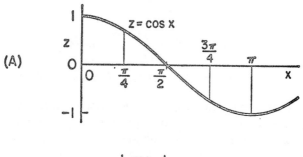

(A)

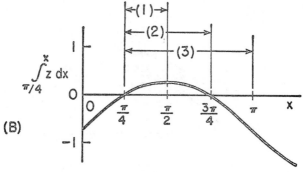

(B)

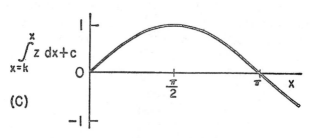

(C)

Fig. 25 (repeated)

Since a and b are numbers with no restrictions placed upon them, they can be changed to any other numbers without invalidating Equation 17. Using b for the a value and a for the b value produces

$$y_a - y_b = \int_{x=b}^{x=a} (dy/dx)\, dx$$

or

$$y_b - y_a = - \int_{x=b}^{x=a} (dy/dx) \, dx \qquad (37)$$

A comparison of Equations 17 and 37 proves Rule 4.

AREA COMPUTATION BY INTEGRATION. Because integration is basically a summation process, it is a useful tool in calculating total areas or volumes when information about their components is known.

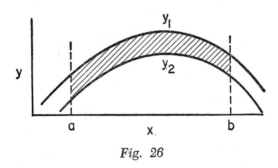

Fig. 26

As an example of area computation, consider the plot of a river shown in Fig. 26, particularly the area of the river shown shaded between $x = a$ and $x = b$. This area is the difference between the area under curve y_1 and the area under curve y_2 between $x = a$ and $x = b$, or stated in calculus notation:

$$\text{Area of river} = \int_{x=a}^{x=b} y_1 \, dx - \int_{x=a}^{x=b} y_2 \, dx$$

Another way of solving the problem exists if y_1 and y_2 are not known as functions of x but the width of the river, $y_1 - y_2$, is known. In this case

$$\text{Area of river} = \int_{x=a}^{x=b} (y_1 - y_2) \, dx \qquad (38)$$

However, the width $(y_1 - y_2)$ must be measured along lines parallel to the y axis *not* along lines perpendicular to the river banks. The reason that the method of measurement is critical can be shown by converting the integral expression of Equation 38 back into a summation expression:

$$\int_{x=a}^{x=b} (y_1 - y_2)\, dx = \lim_{(\Delta x \to 0)} \left[\sum_{x=a}^{x=b} (y_1 - y_2)\, \Delta x \right]$$

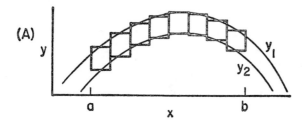

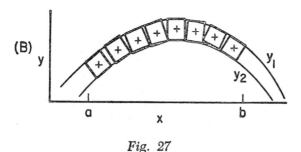

Fig. 27

If Δx is taken to be fairly large, the above summation looks like Fig. 27A and the total area of the rectangles forms a fair approximation to the area of the stream.

Figure 27B shows the result of summing the width of the river (measured normal to the bank) times Δx. Notice the gaps that develop between adjacent rectangles. This alternate summation is a good approximation to the river area only if the river is parallel to the x axis.

An example of area calculation is illustrated in Fig. 28. The problem is to compute the shaded area bounded by curves y_1, y_2 and y_3. The shaded area of Fig. 28 can be considered to be the sum of all the incremental areas that are rectangles h units high and dx units wide between the limits of $x = 1$ and $x = x_d$. Putting this statement in the form of an integral produces:

$$\text{Area} = \int_{x=1}^{x=x_d} h \, dx \qquad (39)$$

Since $h = y_1 - y_2$ between the limits $x = 1$ and $x = 5$ and changes to $h = y_3 - y_2$ between the limits $x = 5$ and

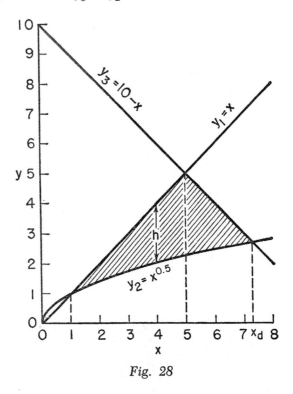

Fig. 28

$x = x_d$, Equation 39 can be rewritten as the sum of two integrals:

$$\text{Area} = \int_{x=1}^{x=5} (y_1 - y_2) \, dx + \int_{x=5}^{x=x_d} (y_3 - y_2) \, dx \qquad (40)$$

However, $y_3 = 10 - x$, $y_2 = x^{0.5}$ and $y_1 = x$. Inserting these relationships into Equation 40 yields:

$$\text{Area} = \int_{x=1}^{x=5} (x - x^{0.5}) \, dx + \int_{x=5}^{x=x_d} (10 - x - x^{0.5}) \, dx \qquad (41)$$

This relationship is ready for integration if the number value of x_d is found. This can be done by solving the y_2 and y_3 expressions for x_d simultaneously since x_d is the value of x that is common to curves y_2 and y_3. The procedure is outlined below.

$$y_2 = y_3$$

$$x_d{}^{0.5} = 10 - x_d$$

$$x_d = (10 - x_d)^2 = 100 - 20x_d + x_d{}^2$$

or

$$x_d{}^2 - 21x_d + 100 = 0$$

By using the quadratic formula,*

$$x_d = 7.3$$

Inserting this value of x_d into Equation 41 as a limit generates:

$$\text{Area} = \int_{x=1}^{x=5} (x - x^{0.5})\, dx + \int_{x=5}^{x=7.3} (10 - x - x^{0.5})\, dx \quad (42)$$

By using Table III, Equation 42 can be evaluated as follows:

$$\text{Area} = \left[\frac{x^2}{2} - \frac{x^{1.5}}{1.5} \right]_{x=1}^{x=5} + \left[10x - \frac{x^2}{2} - \frac{x^{1.5}}{1.5} \right]_{x=5}^{x=7.3}$$

$$\text{Area} = \frac{25}{2} - \frac{5^{1.5}}{1.5} - \left(\frac{1}{2} - \frac{1}{1.5} \right) +$$

$$\left(73 - \frac{(7.3)^2}{2} - \frac{(7.3)^{1.5}}{1.5} \right) - \left(50 - \frac{25}{2} - \frac{5^{1.5}}{1.5} \right)$$

$$\text{Area} = 8.5 \text{ square units}$$

It is suggested that the reader evaluate Equation 42 as an exercise by first separating Equation 42 into five separate integrals using Rule 2 of Table IV; then combining similar integrals using Rule 3 of Table IV, and finally evaluating the

---

*The quadratic formula states: If $ax^2 + bx + c = 0$, then

$$x = \frac{-b \pm \sqrt{b^2 - 4ac}}{2a}.$$

resultant integrals. The answer should be the same as above, 8.5 square units.

MOMENTS. Another example of the use of integration is in the calculation of moments. In this case the problem is to find the moment of inertia of a shaft of density p, length l, and radius r, about its own axis (see Fig. 29). The moment of inertia of a particle of mass M about an axis is

$$I = MR^2$$

where

I = the inertia
M = the mass
R = the distance from the mass to the axis

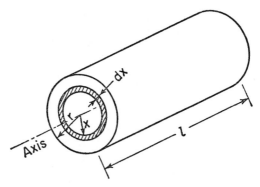

Density = mass/unit volume = ρ

Fig. 29

The key to the problem is that the moment of inertia of a group of masses is equal to the sum of their individual moments of inertia. By dividing the shaft up into thin concentric cylindrical shells, one can find the moment of inertia of each shell and then add their moments of inertia to obtain the moment of inertia of the shaft. A typical shell, shown shaded in Fig. 29, can be thought of as a sheet of material bent into cylindrical form. The area of the shell is its circumference times its length or

$$\text{Shell Area} = 2\pi x l$$

where x is the distance from the axis to the shell and l is the length of the cylinder. The volume of the shell is given by

$$\text{Shell Volume} = \text{Area} \times \text{thickness} = (2\pi x l) \, dx$$

The mass of the shell is the volume times the density, so,

$$\text{Shell Mass} = (2\pi x l p) \, dx$$

The moment of inertia of the shell is the mass times the square of the distance from the axis. As this distance squared is x^2 for every part of the shell,

$$I_{\text{shell}} = (2\pi x l p) \, dx \, (x)^2 = (2\pi x^3 l p) \, dx$$

Adding up the moments of inertia of the shells from the center to the outside results in:

$$I_{\text{shaft}} = \int_{x=0}^{x=r} 2\pi l p x^3 \, dx$$

Removing constants from under the integral sign and evaluating,

$$I_{\text{shaft}} = 2\pi l p \int_{x=0}^{x=r} x^3 \, dx = 2\pi l p \left[\frac{x^4}{4} \right]_{x=0}^{x=r} = \frac{\pi l p r^4}{2}$$

Thus by integration a formula has been developed that permits the rapid calculation of the moment of inertia of any shaft made of any homogeneous material.

4. MULTIPLE INTEGRATION

DOUBLE INTEGRATION. In the previous chapter integration was used to compute the size of an area. Another

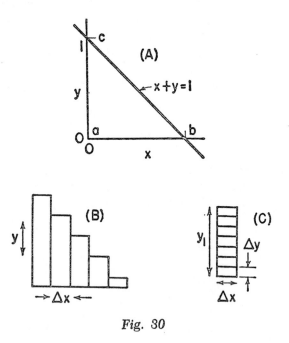

Fig. 30

means of calculating the size of an area is by double integration. This is illustrated by the diagrams in Fig. 30.

The area *a-b-c* under the line in Fig. 30A can be divided into rectangles as before to give the area approximation of Fig. 30B, for which

$$\text{Area} = \sum_{x=a}^{x=b} y \, \Delta x$$

The individual rectangles, $y \, \Delta x$, can also be defined as the sum of several smaller rectangles as shown in Fig. 30C, and defined by

$$y \, \Delta x = \sum_{y=0}^{y=y_1} \Delta y \, \Delta x$$

Combining the last two equations into a single equation results in

$$\text{Area} = \sum_{x=a}^{x=b} \left(\sum_{y=0}^{y=y_1} \Delta y \, \Delta x \right) \tag{43}$$

Equation 43 symbolizes adding the areas $\Delta y \, \Delta x$ together to get a strip and then adding the strips together to get the total area. During the bracketed summation the quantity Δx is fixed, so Equation 43 can be rewritten as

$$\text{Area} = \sum_{x=a}^{x=b} \left(\sum_{y=0}^{y=y_1} \Delta y \right) \Delta x$$

The limit of these summations as Δx and Δy approach the limit of zero is the integral expression:

$$\text{Area} = \int_{x=a}^{x=b} \left[\int_{y=0}^{=y_1} dy \right] dx = \int_{x=a}^{x=b} \int_{y=0}^{y=y_1} dy \, dx \tag{44}$$

The right-hand version of Equation 44 without brackets is the customary mathematical form. However the brackets are still implied and it is important to remember that during the evaluation of the bracketed or "inner" integral, x is a constant.

Evaluating the inner integral of Equation 44 leads to:

$$\int_{y=0}^{y=y_1} dy = [y]_0^{y_1} = y_1 - 0 \tag{45}$$

But y_1 is a general value of y that varies with x and must be stated in terms of x in order to perform the outer integration with respect to x.

From the equation of the curve in Fig. 30,

$$y_1 = 1 - x$$

Substituting this value for the inner integral, the original double integral expression of Equation 44 reduces to a single integral

$$\text{Area} = \int_{x=a}^{x=b} (1 - x) \, dx$$

Applying Rule 2 of Table IV gives:

$$\text{Area} = \int_{x=a}^{x=b} (1) \, dx + \int_{x=a}^{x=b} (-x) \, dx$$

$$= [x]_{x=a}^{x=b} + \left[\frac{-x^2}{2} \right]_{x=a}^{x=b}$$

and as $a = 0$ and $b = 1$,

$$\text{Area} = [x]_0^1 + \left[\frac{-x^2}{2} \right]_0^1$$

$$= (1 - 0) + \left(\frac{-1}{2} - \frac{0}{2} \right) = 1 - \frac{1}{2} = \frac{1}{2}$$

Calculus is not usually applied to such simple problems but the same principles are used to solve involved problems.

MULTIPLE INTEGRALS. A typical multiple integral can be written:

$$M = \iiint q \, dx \, dy \, dz \tag{46}$$

Rewritten with its implied brackets and limits clearly stated the above equation becomes

$$M = \int_{z=a}^{z=b} \left\{ \int_{y=f_3(z)}^{y=f_4(z)} \left[\int_{x=f_1(y,z)}^{x=f_2(y,z)} q \, dx \right] dy \right\} dz \tag{47}$$

In this enlarged notation $x = f_2(y, z)$ means that x is a function of y and z. Instead of integrating the inside brackets between number values, one integrates them between function values such as was done in the first integration of the previous example (Equation 45). In Equations 46 and 47, q is a function of x, y, and z.

As is common with bracketed terms in algebra, *evaluation is done from the inside out.*

The first step in evaluating a multiple integral such as Equation 47 therefore is to evaluate

$$\int_{x=f_1(y,z)}^{x=f_2(y,z)} q \, dx = R$$

In this step the expression in q is written as a function of x, y, and z but y and z are assumed to be constant during the integration. Therefore q is a function of x alone during the integration. However, R is the value of the integral after inserting all limits and R is an expression which contains no x terms. If R turns out to be a function of x, a mistake has been made because the limit values inserted to obtain R should be functions of y and z only, not x.

The second step consists of solving

$$\int_{y=f_3(z)}^{y=f_4(z)} R \, dy = S$$

by evaluating the bracketed integral while assuming that z is constant. Inserting the proper limits creates a new function, S, which is a function only of z.

The third and final integration step,

$$M = \int_{z=a}^{z=b} S \, dx$$

yields a number value for M if a and b are numbers.

MULTIPLE INTEGRATION EXAMPLE. An example may help to provide a visualization of the process of multiple integration. Assume that it is desired to know the total mass of particles in a box if the particles are distributed in an uneven but known manner.

Figure 31 shows the box, the arrangement of reference axes and a sketch of the sequence of the summations that will be used to solve the problem. The stated facts are:

The box is bounded by the planes of $x = 0$, $x = 11$,

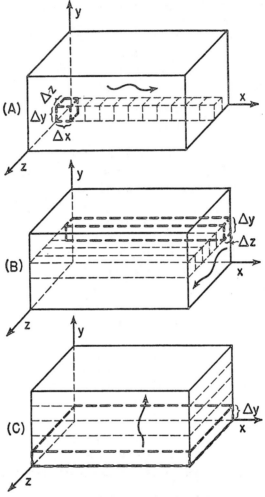

Fig. 31

$y = 0$, $y = 5$, $z = 0$, and $z = 6$. (All stated distances are in centimeters.) The density of material (p) in milligrams per cubic centimeter is

$$p = 0.4 + .003(12 - y)(e^{0.2x})(z^2) \qquad (48)$$

which indicates that the density starts at $(0, 0, 0)$ with a value of 0.4 milligrams per cubic centimeter and increases rapidly in the x^+ and z^+ directions but decreases in the y^+ direction.

In solving this problem a different basic approach will be used than was used in the double integration example. The multiple integral will be constructed from the inside-out instead of from the outside-in as before. The evaluation of the integral will be done from the inside-out as before.

The mass of the $(\Delta x) \cdot (\Delta y) \cdot (\Delta z)$ cube of Fig. 31A is the density at that point in space times the volume of the cube or

$$M_c = \text{cube mass} = p(\Delta x)\,(\Delta y)\,(\Delta z) \qquad (49)$$

where p is the density as given by Equation 48.

A line of cubes can be added by summing in the direction of the wavy arrow of Fig. 31A to find the mass of a bar such as the bar shown heavily dashed in Fig. 31B. In symbol form:

$$M_b = \text{bar mass} = \sum_{x=0}^{x=11} \left(\frac{M_c}{\Delta x}\right)\Delta x = \sum_{x=0}^{x=11} p\,\Delta y\,\Delta z\,\Delta x \qquad (50)$$

Notice that in Equation 50 the parenthetic term is not the mass of the basic cube but the cube mass *per unit length in the direction of integration*. The limits of the summation of Equation 50 are the ends of the box.

A line of bars can now be added by summing in the direction of the wavy arrow of Fig. 31B to find the mass of a slab such as the slab shown heavily dashed in Fig. 31C.

Now in symbol form:

$$M_s = \text{slab mass} = \sum_{z=0}^{z=6} \left(\frac{M_b}{\Delta z}\right)\Delta z \qquad (51)$$

Again the parenthetic term in the summation is not mass but mass per unit length in the direction of integration and the limits of summation are the ends of the box. Substituting Equation 50 for M_b in Equation 51 results in:

$$M_s = \sum_{z=0}^{z=6} \sum_{x=0}^{x=11} p\,\Delta y\,\Delta x\,\Delta z \qquad (52)$$

The final summation is to add slabs in the direction of the wavy arrow of Fig. 31C to find the mass of the box

$$M_b = \text{box mass} = \sum_{y=0}^{y=5} \left(\frac{M_s}{\Delta y}\right) \Delta y$$

$$= \sum_{y=0}^{y=5} \sum_{z=0}^{z=6} \sum_{x=0}^{x=11} p \, \Delta x \, \Delta z \, \Delta y \quad (53)$$

Substituting the analytic expression for p (Equation 48) into Equation 53 and converting Equation 53 from an approximate summation to an exact integral produces the final integral in a form ready for evaluation:

$$M_b = \int_{y=0}^{y=5} \int_{z=0}^{z=6} \int_{x=0}^{x=11} [.04 + $$
$$.003(12 - y)(e^{0.2x})z^2] \, dx \, dz \, dy \quad (54)$$

Notice that the steps followed from Equation 49 to Equation 53 automatically rearranged the Δx, Δy, and Δz terms into proper order for integration.

The first step in the evaluation of Equation 54 is to evaluate the innermost integral. This step corresponds to the diagram of Fig. 31A and since all summation occurs at the same value of y and z, they are treated as constants during the integration.

$$\int_{x=0}^{x=11} [.04 + .003(12 - y)(e^{0.2x})(z^2)] \, dx$$

$$= [.04 + .003(12 - y)z^2] \int_{x=0}^{x=11} (e^{0.2x}) \, dx$$

$$= [.04 + .003(12 - y)z^2][5e^{0.2x}]_{x=0}^{x=11}$$

$$= [.04 + .003(12 - y)z^2][45 - 5]$$

$$= 1.6 + .120(12 - y)z^2 \quad (55)$$

The second step in the evaluation of Equation 54 is to insert Equation 55 into Equation 54 as follows:

$$M_b = \int_{y=0}^{y=5} \int_{z=0}^{z=6} [1.6 + .120(12 - y)z^2] \, dz \, dy \quad (56)$$

Notice that x has completely vanished from the equation for M_b. Again the inner summation is performed with all variables except the variable of integration (z this time) treated as constants. This integration is diagramed in Fig. 31B.

$$\int_{z=0}^{z=6} [1.6 + .120(12 - y)z^2]\, dz$$

$$= [1.6z]_{z=0}^{z=6} + .120(12 - y) \left[\frac{z^3}{3}\right]_{z=0}^{z=6}$$

$$= 113 - 8.6y \tag{57}$$

The final integration is at hand if Equation 57 is inserted into Equation 56

$$M_b = \int_{y=0}^{y=5} (113 - 8.6y)\, dy = 457.5 \text{ milligram} = 0.4575 \text{ grams}$$

The reader is invited to check the last calculation as an exercise. The diagram for this last integration is shown in Fig. 31C.

The triple and double integrals are not the only possible multiple integrals. With sufficient courage there is no limit!

DIMENSIONS. The dimensional units of a multiple integral are the product of the units of the quantity being integrated and the units of the variables of integration. In the example just completed, [mass per unit volume] was integrated by three [length] variables to produce a dimensional product of [mass].

III

DIFFERENTIAL EQUATIONS I

(Time Calculus)

5. EQUATIONS OF MOTION

EQUATIONS OF MOTION. Equations of motion are equations in which time is the independent variable. A major use of calculus principles is to devise and solve equations that describe motion among parts of a machine or an electrical circuit; therefore it is important for a student of calculus to understand equations of motion and their components. In addition, since Laplace operational calculus is usually derived in terms of motion equations, a familiarity with equations of motion is a valuable preface to Chapter 10.

TIME DERIVATIVES. An example of a variable that is a function of time is shown in the plot in Fig. 32A of the distance between a train and its roundhouse during a day's run as a function of time. Notice that at the end of the day the train hasn't returned to the roundhouse.

The first derivative of distance with respect to time is called "speed" or "velocity."* In Fig. 32A the velocity at any time can be found by measuring the slope of the curve since the first derivative, Δ distance $\div$ Δ time (as Δ time approaches zero), is the slope of the curve. Figure 32B shows a plot of the velocity of the curve of Fig. 32A. The velocity is zero at points a, b, c, is positive at d and is negative at e. As velocity is the limit of the ratio of a distance divided by time, its dimensions are [miles/hour].

The second derivative of distance with respect to time (the first derivative of velocity) is called acceleration,

* Speed and velocity are not strictly the same. In motion problems the proper term is speed, but unfortunately most engineers and physicists use the term velocity; so this book will also.

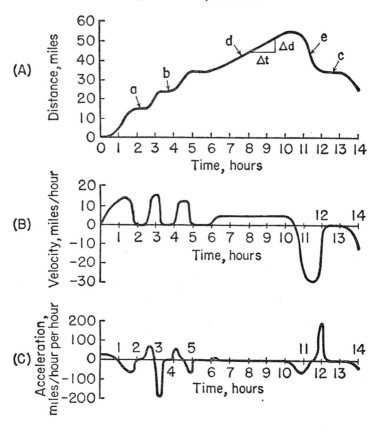

Fig. 32

which is the ratio of changes of velocity to changes in time. In Fig. 32C the acceleration of the train is plotted against time. Notice that when the train is stopped, as at points *a*, *b*, and *c*, the velocity and acceleration are both zero. This is because when the train is at rest it has no velocity and its velocity is not changing. Notice also that a positive displacement slope means a positive velocity and a concave upward section of the displacement curve (one in which velocity increases as time increases) means a positive acceleration. At points where the acceleration is zero, the velocity is constant, and the displacement curve is a straight line. As practice directed toward developing a sense of the relationship among displacement, velocity, and acceleration, it is suggested that the reader try his hand at drawing Figs. 32B and 32C freehand while looking only at Fig. 32A.

The next higher time derivative in the displacement-velocity-acceleration series is "jerk." There are many time derivatives of use in solving problems in physics and engineering. For example, power is the first time derivative of energy, force is the first time derivative of momentum (in the case of a moving mass), and current is the first time derivative of charge (in an electric circuit).

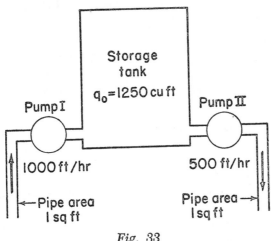

Fig. 33

TIME INTEGRALS. As an example of the use of time-integral expressions, assume that a record has been kept of the rate of flow in and out of a storage tank, Fig. 33, and it is desired to know the total quantity of fluid in the tank at the end of a ten-hour period.

Two pumps are shown. Pump I, when turned on, pumps water into the tank at a velocity of 1000 ft per hr. Pump II, when operated, pumps water out of the tank at a velocity of 500 ft per hr. If both inlet and outlet pipes have the same diameter, 1 sq ft, the net result of both pumps operating at the same time is to fill the tank at half the rate of pump I working alone.

The integral expression that describes Fig. 33 is:

$$q = k \int v \, dt \qquad (58)$$

where q is the total fluid in the tank, v is the velocity of the fluid flow (positive if filling the tank and negative if emptying the tank). The constant, k, must be evaluated from the physics of the problem, as follows:

Assume that v has been constant for a short period, say $v_1 = 1000$ ft/hr for a period of 1 hr. This would cause q to change by 1000 cu ft as the pipe area was stated to be 1 sq ft. Rewriting the integral expression, Equation 58, for this case in complete form gives:

$$q = k \int_{t=0}^{t=1\ \text{hr}} v_1 \, dt = k[v_1 t]_{t=0}^{t=1\ \text{hr}}$$

$$= k[(1000\ \text{ft/hr})(1\ \text{hr} - 0)] = k(1000\ \text{ft})$$

or

$$1000\ \text{cu ft} = k(1000\ \text{ft})$$

therefore

$$k = 1\ \text{sq ft}$$

If v is known as a function of time, q can be computed from

$$q_a = k \int_{t=0}^{t=a} v \, dt + q_0$$

where q_0 is the value of q at time $t = 0$ (the initial condition) and q_a is the value of q at time $t = a$. Another way of stating the q vs v relationship is to say that q is changed from its initial value by an amount proportional to the area under the curve of v vs time. The latter approach will be used in the following example.

Figure 34A shows a schedule of pump operation. Pump I is on for the 1st, 4th, and 5th hours and pump II is on for the 2nd, 3rd, 5th, and succeeding hours. Figure 34B shows the resultant velocity of flow into the tank. A negative velocity (when only pump II is working) means that the tank is being emptied. Figure 34C shows the resultant q

with q_0 given as 1250 cu ft. Notice that q increases whenever v is positive and that the rate of increase is proportional to v. Figure 34D shows what would have happened to the

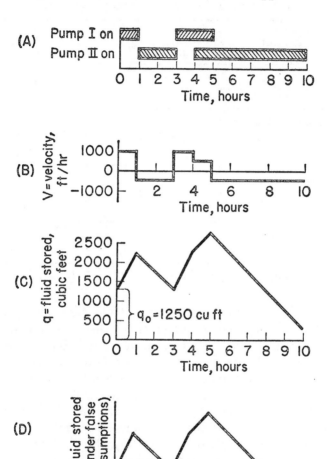

Fig. 34

plot of q vs time if q_0 had been forgotten. The result would have been a tank holding a negative quantity of fluid for the last two hours!

TABLE V

COMMONLY USED SYMBOLS FOR DIFFERENTIATION
AND INTEGRATION WITH RESPECT TO TIME

Traditional form	x	$\dfrac{dx}{dt}$	$\dfrac{d^2x}{dt^2}$	$\dfrac{d^3x}{dt^3}$	$\displaystyle\int x\,dt$
Dot notation	x	$\dot{x}$	$\ddot{x}$	$\dddot{x}$	$\displaystyle\int x\,dt$
Heaviside operational notation	x	Px	P^2x	P^3x	$(1/P)x$ or $P^{-1}x$
Alternate operational notation	x	Sx	S^2x	S^3x	x/S or $S^{-1}x$
Advanced calculus notation	x	$D(x)$	$D^2(x)$	$D^3(x)$	$D^{-1}(x)$
Function notation	$f(t)$	$f'(t)$	$f''(t)$	$f'''(t)$	$f^{-1}(t)$

SHORTHAND NOTATIONS. Differentiation and integration
with respect to time are often abbreviated and the more
commonly used symbols are listed in Table V for reference.
The traditional form is always in good taste. The dot notation is very good except that an occasional fly spec will add
errors. It was devised because it is easy to typewrite. The
operational notations have the disadvantage of sometimes
being mistaken for algebraic expressions but are in common
use, especially the Heaviside notation. The last two notations are usually found in advanced mathematics books.

WRITING MOTION EQUATIONS. The writing of equations
of motion can be explained by a mechanical and an electrical example. It is suggested that the reader glance through
each example and then reread both examples in detail.

EXAMPLE 1: Figure 35 shows a simple suspension system where K and C are a spring and a damper respectively
that support the mass, M. Desired is a plot of x, the vertical

motion of the mass with respect to its rest position, as an applied force, F_y, varies as a function of time. The equations of motion can be written by several methods but the method of force balance is most common. By "force balance"

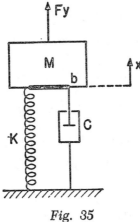

Fig. 35

is meant that the sum of all forces acting on point b on the mass is zero if the forces of inertial reaction are included.*

Calling displacement, velocity, acceleration, and force in an upward direction positive, leads to the following statements:

Force upward on b due to applied force, F_y, plus
Force upward on b due to mass inertial reaction, F_1, plus
Force upward on b due to spring; F_2, plus
Force upward on b due to viscous damper, F_3, equals zero,

or

$$F_y + F_1 + F_2 + F_3 = 0 \qquad (59)$$

The values of F_1, F_2, and F_3 can be converted to func-

* An excellent reference on the writing of motion equations is "Mathematical and Physical Principles of Engineering Analysis" by Walter C. Johnson, McGraw-Hill Book Co., New York, 1944.

tions of x by the following relationships:

$$F_1 = -M\ddot{x} \qquad (60)$$

where x = vertical acceleration of point b (or any other point on mass M);

$$F_2 = -Kx \qquad (61)$$

where K = spring constant (assuming a perfectly elastic linear spring);

$$F_3 = -C\dot{x} \qquad (62)$$

where C = damping constant (assuming a linear viscous damper).

Substituting Equations 60 to 62 in Equation 59 leads to:

$$F_y - M\ddot{x} - Kx - C\dot{x} = 0$$

or $\qquad\qquad\qquad\qquad\qquad\qquad\qquad\qquad (63)$

$$M\ddot{x} + C\dot{x} + Kx = F_y$$

Since F_y is a function of time, Equation 63 is a relationship between x and time. In the next section this equation will be interpreted.

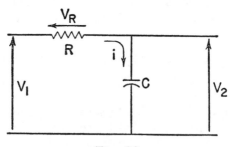

Fig. 36

EXAMPLE 2: In the circuit of Fig. 36, V_1 is an applied voltage and V_2 is the output voltage sought. If V_1 is a function of time, V_2 will be also. The "motion" in this circuit is that of electric charges. It is equally proper, however, to think of a voltage as having "motion," meaning "change of value with changes in time." In Fig. 36, the voltage across the capacitor is

$$V_2 = \frac{1}{C} \int_0^t i \, dt + V_2(0) \tag{64}$$

where
C = the capacity of the capacitor
i = the loop current
$V_2(0)$ = the value of V_2 at time $t = 0$

and the voltage across the resistor is given by:

$$V_r = iR$$

where R = the resistance of the resistor.

By Kirchhoff's voltage law,* the sum of the voltages around a closed loop equals zero. Therefore:

$$V_1 - V_r - V_2 = 0$$

Substituting iR for V_r gives

$$V_1 - iR - V_2 = 0 \tag{65}$$

If two expressions are equal, their integrals or derivatives are also equal. Therefore if an equation is integrated or differentiated, term by term, the resulting expression is also an equation as valid as the first. Differentiating Equation 64 results in:

$$\dot{V}_2 = \frac{1}{C} i$$

or

$$i = C\dot{V}_2$$

which, when substituted in the second term of Equation 65, produces:

$$V_1 - C\dot{V}_2R - V_2 = 0$$

or

$$CR\dot{V}_2 + V_2 = V_1 \tag{66}$$

Since V_1 is a function of time, this is an equation of motion for V_2 as it defines V_2 as a function of time.

* See any electrical engineering text or handbook for a discussion of Kirchhoff's laws.

INTERPRETATION OF MOTION EQUATIONS. A motion equation can be interpreted with ease if each term is understood.

EXAMPLE 1: The motion equation of the suspension system of Fig. 35,

$$M\ddot{x} + C\dot{x} + Kx = F_y \tag{63}$$

merely expresses a relationship that must be satisfied among the number values of x, its derivatives and F_y. The "solution" of the equation, a curve of x as a function of time, will depend on also knowing the boundary conditions. Almost without exception these conditions in motion equations are given at time $t = 0$ or $t = 0^+$.* As this is a second-order differential equation, two initial conditions are needed.†

In this example it will be assumed that $\dot{x} = x = 0$ at $t = 0$ and that F_y is a constant positive force, F_k, applied at $t = 0^+$ and continued indefinitely (a step input). At $t = 0^+$, Equation 63 becomes:

$$M\ddot{x} + C(0) + K(0) = F_k$$

or

$$M\ddot{x} + 0 + 0 = F_k$$

Therefore

$$\ddot{x} = F_k/M \quad (\text{at } t = 0^+)$$

Putting these bits of information together it is apparent that since $x = 0$ at time $t = 0^+$, the curve of x vs time will start at the origin. Furthermore $\dot{x} = 0$ at $t = 0^+$, so that the slope of x vs time will be zero or parallel to the x axis at $t = 0^+$. In addition, since the acceleration, $\ddot{x}$, is a positive number at $t = 0^+$, the curve of x vs time is concave upward to start. Thus the start of the solution, x vs time, can be drawn as shown in Fig. 37 by line o-a.

At time $t = \infty$, any physical system will either oscillate or reach a steady-state condition such as a constant displacement or a constant velocity. In this example it will be assumed from experience with systems of this type that as $t \to \infty$,

---

* 0^+ means an infinitesimal time after time zero. This notation for initial values is often used to avoid confusion in the case of functions that change abruptly at time $t = 0$.

† The order of equations is discussed in Chapter 13.

all motion will cease. With a constant force applied to the system, energy will be dissipated in the damper until velocity ceases. If velocity = 0 over a period of time, its first derivative, acceleration, must also equal zero. Therefore the final

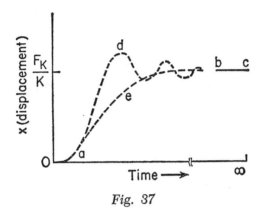

Fig. 37

values of $\dot{x}$ and x are zero. Inserting these in Equation 63 gives

$$0 + 0 + Kx = F_k \quad (\text{at } t = \infty)$$

or

$$x = \frac{F_k}{K} \quad (\text{at } t = \infty)$$

which determines segment *b-c* in Fig. 37.

The shape of the curve between points *a* and *b* cannot be determined without inserting specific number values into Equation 63 for *C*, *K*, and *M*. Dotted in Fig. 37 are two of a family of possible curves. Curve *d* represents a case in which *C* is small compared to *K* and curve *e* is typical of cases in which *C* is large compared to *K*. Methods of solution of time differential equations to determine intermediate values of $\dot{x}$ will be discussed in Chapter 6.

EXAMPLE 2: The motion equation of the *CR* circuit of Fig. 36 is:

$$CRV_2 + V_2 = V_1 \tag{66}$$

Assume that initial conditions are stated as $V_1 = 0$ and $V_2 = 1$ at $t = 0^+$. Physically this means that V_1 is short-circuited at $t = 0^+$ and that the capacitor is initially charged. Inserting the initial conditions reduces Equation 66 to

$$CR\dot{V}_2 + 1 = 0 \qquad (\text{at } t = 0^+)$$

or

$$\dot{V}_2 = -\frac{1}{CR} \quad (\text{at } t = 0^+)$$

Summarizing, at $t = 0^+$, V_2 has a value of 1 and is decreasing ($\dot{V}_2$ is negative) at a slope of $-1/CR$. This is sketched in as segment *a-b* in Fig. 38.

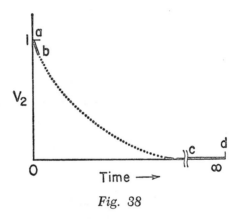

Fig. 38

As time becomes infinite, if it is assumed that the circuit will not oscillate, V_2 will be zero so that

$$0 + V_2 = 0 \quad (\text{at } t = \infty)$$

or

$$V_2 = 0 \quad (\text{at } t = \infty)$$

which determines segment *c-d* in Fig. 38. What happens in between is shown dotted, although nothing has been stated yet that would assure one that the dotted curve does not drop below the time axis. For this intermediate data, it is necessary to solve the equation of motion.

SUMMARY. In summary, it is possible by inspection of a motion equation, when given the initial conditions, to determine the initial shape of the curve of the solution. If there is good reason from physical information to believe that the system will not oscillate, the final value can be determined. *Intermediate values require solving the equation.*

6. SOLVING EQUATIONS OF MOTION

SOLVING MOTION EQUATIONS. Of the many methods commonly used to solve motion equations, four of them will be discussed by using each to solve the equation of the electric resistance-capacitance circuit of Fig. 36 for the case of a particular input signal. These four have been selected to give the reader an appreciation of the process and enough familiarity to remove any mystic aura. Solving a differential equation (or any equation for that matter) means manipulating the equation (or group of equations) to present the same information in more readily usable or convenient form. Thus a time differential equation in V_0 gives V_0 as a function of time, but in an awkward form. After solving the equation, the same information is obtained in the form of a graph of V_0 vs time or an algebraic expression of V_0 as a function of time.

METHOD 1: STEP-BY-STEP NUMERICAL SOLUTION. This method has many forms and refinements and is the basis of the solution of time differential equations by digital computers and digital differential analyzers. It is an approximate method but by reducing the size of the increment of time chosen, it can be made as accurate as desired.

Assume that the circuit of Fig. 36 has a ramp voltage input for V_1 as shown in Fig. 39. Expressed analytically,

$$V_1 = 2t$$

As before, the motion equation for the circuit of Fig. 36 is

$$CR\dot{V}_2 + V_2 = V_1 \tag{66}$$

Assume that in this case $CR = 4$.
Equation 66 now becomes

$$4\dot{V}_2 + V_2 = 2t \qquad (67)$$

If it is assumed that the capacitor is initially uncharged, the

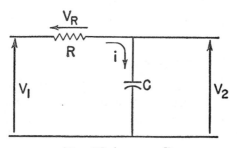

Fig. 36 (*repeated*)

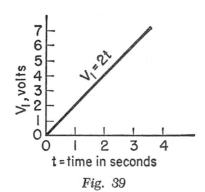

Fig. 39

initial conditions become

$$V_2 = 0 \quad \text{and} \quad \dot{V}_2 = 0 \quad (\text{at } t = 0^+) \qquad (68)$$

Equations 67 and 68 will now be solved for V_2 as a function of time.

An approximate form of Equation 67 is

$$\frac{4\,\Delta V_2}{\Delta t} + V_2 = 2t \qquad (69)$$

In this case ΔV_2 is the change in V_2 that corresponds to a change in t of Δt. Rearranging Equation 69:

$$4\,\Delta V_2 = (2t - V_2)\,\Delta t$$

or

$$\Delta V_2 = \left(\frac{t}{2} - \frac{V_2}{4}\right)\Delta t \qquad (70)$$

Assume a standard value for Δt of 0.1 seconds. This changes Equation 70 to:

$$\Delta V_2 = 0.05t - 0.025V_2 \qquad (71)$$

in which ΔV_2 is the change in V_2 during any 0.1 second time interval. If the further approximation is made that V_2 and t in Equation 71 are the values at the beginning of each 0.1 second interval, Equation 71 can be used directly to compute the change in V_2 for each 0.1 second interval as follows:

$$\underset{(0<t<0.1)}{\Delta V_2} = 0.05(0) - 0.025(0) = 0$$

giving

$$\underset{(t=0.1)}{V_2} = \underset{(t=0)}{V_2} + \underset{(0<t<0.1)}{\Delta V_2} = 0 + 0 = 0$$

Repeating the process:

$$\underset{(0.1<t<0.2)}{\Delta V_2} = 0.05(0.1) - 0.025(0) = 0.005$$

giving

$$\underset{(t=0.2)}{V_2} = \underset{(t=0.1)}{V_2} + \underset{(0.1<t<0.2)}{\Delta V_2} = 0 + 0.005 = 0.005$$

Repeating again:

$$\underset{(0.2<t<0.3)}{\Delta V_2} = 0.05(0.2) - 0.025(0.005)$$

$$= 0.01 - 0.000125 = 0.009875$$

giving

$$\underset{(t=0.3)}{V_2} = \underset{(t=0.2)}{V_2} + \underset{(0.2<t<0.3)}{\Delta V_2} = 0.005 + 0.009875 = 0.014875$$

Continuing this process gives the graph of V_2 as a function of time shown in Fig. 40 as a solid line.

Notice that the curve obtained is a staircase and approximates the correct solution which is the dashed curve. If this problem were solved on a digital computer the steps could be made as small as desired for increased accuracy.

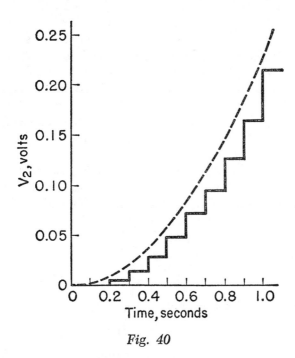

Fig. 40

METHOD 2: CLASSICAL METHOD. This method consists of breaking the original equation into two equations, then using separate techniques to solve each part, finally combining the solutions and using initial conditions to fix values of constants that develop in the process.

As before, the equation to be solved is:

$$4\dot{V}_2 + V_2 = 2t \tag{67}$$

which can be replaced by two equations:

$$4\dot{V}_{2A} + V_{2A} = 2t \tag{72}$$

and

$$4\dot{V}_{2B} + V_{2B} = 0 \tag{73}$$

The procedure is to solve Equation 72 for V_{2A}, solve Equation 73 for V_{2B} and then add these solutions to obtain the required V_2.

Equation 72 appears to be the same as the original equation, Equation 67, but it will be solved only for a particular solution which may not be complete. It is because of this restriction that Equation 72 is simpler to solve than the original equation. The solution of Equation 72 usually corresponds to the steady-state terms in the total solution. V_{2A} is called the particular solution of the original equation.

The equation for V_{2B}, Equation 73, is called the complementary equation and V_{2B} is called the complementary solution of the original equation, Equation 67. V_{2B} usually contains the transient terms of V_2.

To solve for V_{2A}, first notice that V_{2A} or its derivative must have some term which is a number times t. A particular solution is now assumed (a guess) and tried. If

$$V_{2A} = At + B$$

where A and B are constants

then $$\dot{V}_{2A} = A$$

and Equation 72 becomes:

$$4A + (At + B) = 2t \qquad (74)$$

However, if Equation 74 is to be equivalent to Equation 72 for all values of time, then it follows that

$$At = 2t$$

and

$$4A + B = 0$$

from which

$$A = 2 \quad \text{and} \quad B = -8$$

so that

$$V_{2A} = 2t - 8 \qquad (75)$$

is the particular solution sought.

To solve for V_{2B}, the complementary equation, Equation 73 is rewritten as:

$$4\dot{V}_{2B} = -V_{2B} \tag{76}$$

The solution of Equation 76 must be a function of time which, when differentiated, equals itself times a constant. Such a function is:

$$V_{2B} = Ce^{Dt}$$

since

$$\dot{V}_{2B} = DCe^{Dt}$$

where C and D are constants. Substituting this solution for V_{2B} in Equation 76 gives

$$4DCe^{Dt} = -Ce^{Dt}$$

from which

$$4DC = -C$$

and

$$D = -\tfrac{1}{4}$$

therefore

$$V_{2B} = Ce^{-t/4} \tag{77}$$

is the solution of the complementary equation.

To evaluate C in Equation 77 requires applying the initial conditions to the complete solution. To finish the job of solving for V_2, Equations 75 and 77 are summed to yield:

$$V_2 = V_{2A} + V_{2B} = 2t - 8 + Ce^{-t/4} \tag{78}$$

The initial conditions are that $V_2 = 0$ at $t = 0$; therefore at $t = 0$ Equation 78 becomes:

$$V_2 = 0 = 0 - 8 + Ce^0 = -8 + C$$

or

$$0 = C - 8 \quad \text{and} \quad C = 8$$

since e raised to the zero power is 1.

The complete solution is therefore determined by substituting 8 for C in Equation 78 to give:

$$V_2 = 2t - 8 + 8e^{-t/4}$$

This solution is plotted as the dashed line in Fig. 40 so that it can be compared with the previously developed approximate solution.

METHOD 3: USE OF AN ELECTRONIC ANALOG COMPUTER. Involved time differential equations are often solved by general purpose electronic analog computers which are essentially mathematical models of the equation.

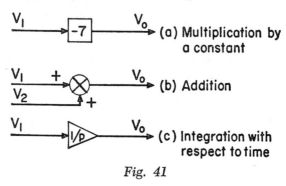

Fig. 41

To solve Equation 67, a few devices are needed that are readily available in an electronic analog computer. Each of these devices has a symbol. Many systems of symbols are in common use, but the symbols shown in Fig. 41 are typical of those used for the initial problem setup on an electronic analog computer.

Symbol (a) is for a device that multiplies a voltage by a constant, in this case —7, so that for the device shown

$$V_0 = -7V_1 \qquad \text{Fig. 41(a)}$$

Symbol (b) is for a device that produces an output voltage, V_0, that is the sum of two input voltages, V_1 and V_2. Its equation is therefore

$$V_0 = V_1 + V_2 \qquad \text{Fig. 41(b)}$$

Symbol (c) is for a device that integrates its input voltage with respect to time, abbreviated as $1/P$ as discussed before. In this case

$$V_0 = \int V_1 \, dt \qquad \text{Fig. 41(c)}$$

Ordinarily there are provisions for the insertion of a constant of integration but that will not occur in this problem.

In each device the input and output signals are voltages which represent variables in the problem being solved. The devices themselves are various forms of very stable high gain d-c amplifiers with negative feedback.

The method usually used to connect these devices together to solve an equation assumes that the highest derivative term in the equation is known and then forces the computer elements to generate the assumed highest derivative.

The equation to be solved, Equation 67, will be used in this form:

$$\dot{V}_2 = \frac{2t - V_2}{4}$$

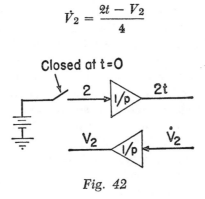

Fig. 42

The first step toward a computer solution is shown in Fig. 42. At the top a method is shown for obtaining a voltage signal that varies as the $2t$ term. This is done by time-integrating a fixed signal of two volts. At the bottom of Fig. 42, V_2 is generated by time-integrating the assumed signal, $\dot{V}_2$.

If $2t$ and V_2 are now combined to form the term $(2t - V_2)/4$, $\dot{V}_2$ will be determined.

In Fig. 43, V_2 is multiplied by (-1), added to $2t$, the sum is multiplied by $\frac{1}{4}$ and the resultant signal is the $\dot{V}_2$ originally assumed.

To solve the equation, the switch is closed causing $2t$ to build up. The V_2 signal is fed to a recording device run

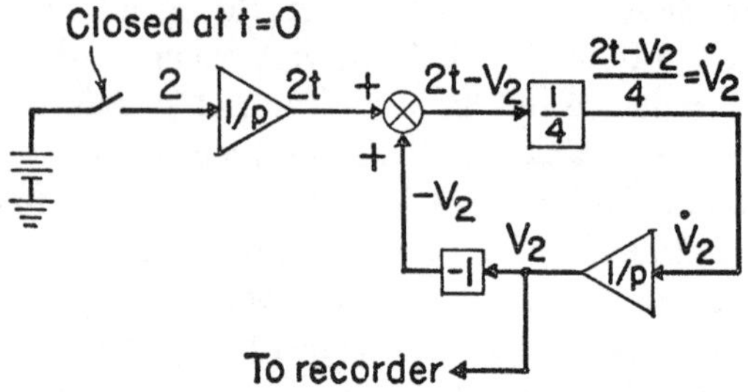

Fig. 43

by a clock motor and the recorder trace is then an accurate plot of V_2 as a function of time.

METHOD 4: LAPLACE OPERATIONAL CALCULUS. Laplace operational calculus is a method of transforming a time differential equation into an equivalent algebraic equation, which can be readily solved. The solution of the algebraic equation is then transformed in reverse (called "inverse transforming") into its corresponding time variable to give the desired solution. This process is not involved but to understand it and see its basis requires a knowledge of complex number theory so this discussion will be limited to a demonstration with a more complete explanation postponed to Chapters 10 and 11.

Table VI is a partial table of transforms containing all the transformations needed for this problem.

In Table VI, p is an algebraic variable and is the variable in the transformed equation that corresponds to t in the original equation. The variable V_2 transforms from a function of time to V_{2p} which is a function of p. The first time derivative of V_2 transforms to p times the transform of V_2 minus the numerical value of V_2 at $t = 0$ (the initial condition of V_2) which is symbolized by $V_2(0)$. Notice that the operation of taking a derivative has transformed into the operation of multiplying by p and subtracting a constant. A number transforms into the same number divided by p. As shown in Table VI, an exponential decay transforms into a simple fraction in p. The same table will be used for both the transforming and inverse transforming processes.

TABLE VI

ELEMENTARY LAPLACE TRANSFORMS

Real Quantity	Transformed Quantity
V_2	V_{2p}
$\dot{V}_2$	$pV_{2p} - V_2(0)$
1	$1/p$
t	$1/p^2$
e^{-at}	$\dfrac{1}{p+a}$

The differential equation and the given intial conditions

$$4\dot{V}_2 + V_2 = 2t \qquad V_2 = 0 \quad (\text{at } t = 0)$$

are transformed term by term into functions of the algebraic variable p to give:

$$4p(V_{2p}) - 0 + V_{2p} = \frac{2}{\mathrm{p}^2}$$

which can be rewritten as:

$$V_{2p} = \frac{2}{p^2(4p + 1)} \tag{79}$$

By inverse transforming Equation 79, the desired solution is obtained. To facilitate this operation Equation 79 is rewritten as

$$V_{2p} = \frac{2}{p^2} + \frac{-8}{p} + \frac{32}{(4p + 1)} \tag{80}$$

The reader can verify the last step by placing all the terms in Equation 80 over a common denominator of $p^2(4p + 1)$.*

To facilitate the inverse transformation, Equation 80 is

* The process used to convert Equation 79 to Equation 80 is outlined in Chapter 11 under the heading "Partial Fractions."

rearranged to read

$$V_{2p} = 2\left(\frac{1}{p2}\right) - 8\left(\frac{1}{p}\right) + 8\left(\frac{1}{p + \frac{1}{4}}\right) \tag{81}$$

Inverse transforming Equation 81 term by term gives:

$$V_2 = 2(t) - 8(1) + 8(e^{-t/4})$$

or

$$V_2 = 2t - 8 + 8e^{-t/4}$$

as derived previously.

IV

COMPLEX VARIABLES

7. COMPLEX ALGEBRA

NUMBERS REDEFINED. An understanding of complex numbers is a prerequisite to the study of Laplace operational calculus. The following are complex numbers:

$$17 + 2j$$

$$3 - j$$

$$1.7 - 2.1j$$

Or, to put it in general terms,

$$a + bj = \text{a complex number}$$

where a and b are ordinary numbers, either positive or negative, integer, fractional or irrational, and j^* is the square root of minus one. That is,

$$j = \sqrt{-1}$$
$$j^2 = \sqrt{-1} \times \sqrt{-1} = (\sqrt{-1})^2 = -1$$

The portion of the complex number represented by a is called the real part and bj is called the imaginary part. The common use of the word "imaginary" for the part of the complex number that includes the square root of -1 is unfortunate, for "imaginary" usually means "unreal"

* Both i and j are used to represent $\sqrt{-1}$ in books and articles. There is a trend toward the use of j because of the prevalent use of i to represent electric current in physics and electrical engineering.

and $\sqrt{-1}$ is no more unreal than -7 or $+3$! All numbers are unreal!! An explanation of this statement follows.

Numbers can be regarded as abstract symbols that have an orderly sequence and can be manipulated according to a set of rules to produce other numbers. By changing the rules, different algebras can be invented. When someone notices that a physical problem can be converted into number symbol form and that by manipulating the numbers within the rules of algebra a number results that can be interpreted as an answer to the problem, then an applied mathematics is born. Addition, subtraction, and multiplication of positive integers lead to the solution of problems of counting and arranging. The introduction of fractions permits problems in proportions to be solved. Negative numbers (what could be more unreal than less than nothing?) complete the kit of mathematical tools to the point that business credit and debit problems can be solved. From integers to fractions to negative numbers to complex numbers—a chain of abstract symbols, all unreal—and the square root of minus one is no more abstract than plus two.

THE NEED FOR COMPLEX NUMBERS. Today there are *uses* for complex numbers in electrical phasor analysis, vibration analysis, in symmetrical component analysis, in network analysis, in servomechanism analysis, and in many other problems. A *need* for complex numbers, however, arises from a simple problem in algebra: find x if $x^2 = -4$. Since the square of either $+2$ or -2 is $+4$, neither can be a solution. If, however, there were a number, j, such that $j^2 = -1$, then the problem could be solved as follows:

$$x^2 = -4$$
$$x^2 = 4(-1)$$
$$x^2 = 4j^2 = (2j)^2$$
$$x = \pm 2j$$

If the symbol $j = \sqrt{-1}$ is added to ordinary algebra, all numbers, positive and negative, have two square roots, three cube roots, etc.

REPRESENTING COMPLEX NUMBERS. Ordinary numbers such as 0, ½, —4, and +6¾ have a single (real) part and can be represented as points on a line. Since a complex number has two parts (real and imaginary), it can be represented as a point on a plane. This is shown in Fig. 44.

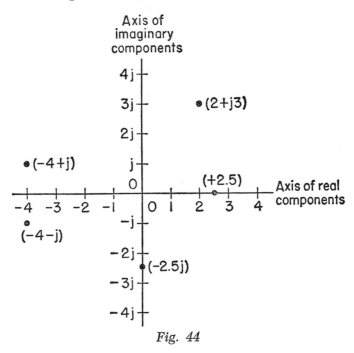

Fig. 44

The abscissa is called the axis of real components and the ordinate is called the axis of imaginary components. Any complex number can be represented as a point in the "complex plane" as this plane is called. Several points are indicated. Notice that any ordinary number (a complex number with an imaginary part equal to zero) lies somewhere on the axis of real components. It is also apparent that for a complex number to equal zero, both its real and imaginary components must equal zero. Real numbers can be considered to be special cases of complex numbers.

ALGEBRAIC OPERATIONS WITH COMPLEX NUMBERS. The addition of two or more complex numbers is accomplished by adding their components together as shown in these two examples:

$$
\begin{array}{ll}
1 + 3j & (a) \\
2 + j & (b) \\
\hline
3 + 4j & (c) \\
(a) + (b) = (c) &
\end{array}
$$

$$
\begin{array}{ll}
-4 - j & (d) \\
4 - j & (e) \\
\hline
-2j & (f) \\
(d) + (e) = (f) &
\end{array}
$$

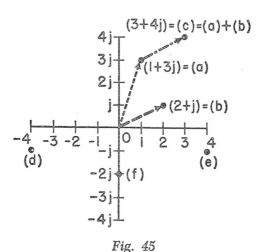

Fig. 45

The first summation is shown in Fig. 45. The addition of $(a) + (b)$ can be interpreted in graphical form by assigning a vector arrow from the origin to each point [the dotted arrow represents (a) and the dashed arrow represents (b)] and then moving one arrow in a parallel motion so that its tail touches the tip of the arrow that was not moved. In the construction shown, the dashed arrow was moved to become the dash-dot arrow. The tip of the moved arrow then falls on the desired sum. It is suggested that the reader practice by performing the graphical construction for $(d) + (e) = (f)$.

Subtraction of complex numbers is the same as addition, with the intermediate step of changing the sign of the subtrahend. Thus this subtraction problem

$$(7 + 3j)$$
$$-(4 - 3j)$$

is converted to an addition problem

$$(7 + 3j)$$
$$(-4 + 3j)$$

to yield:
$$(3 + 6j)$$

Multiplication of complex numbers proceeds in the same manner as ordinary algebraic multiplication. All that need be noted is that $j^2 = -1$. For example:

$$(7 + 2j)(3 - 4j) = (7)(3) + (2j)(3) + (7)(-4j) + (2j)(-4j)$$
$$= 21 + 6j - 28j - 8j^2$$
$$= 21 - 22j - 8(-1)$$
$$= 21 - 22j + 8$$
$$= 29 - 22j$$

A graphical interpretation of complex number multiplication will be given in the next section.

The complex conjugate is an interesting and useful number. The complex conjugate of $(a + jb)$ is $(a - jb)$. In the complex plane the complex conjugate of a number is at the location of its image if the axis of reals is replaced by a mirror. This is illustrated in Fig. 46 on the next page.

The numbers plotted as a and a' are complex conjugates of each other as are b and b', c and c', etc. If two numbers are complex conjugates of each other, their product is a real number. For example:

$$(7 + j3)(7 - j3) = 49 + j21 - j21 - j^2 9$$
$$= 49 + 9 = 58$$

In general:
$$(a + jb)(a - jb) = a^2 + b^2$$

Division of complex numbers can be accomplished by

using multiplication by the complex conjugate of the denominator to reduce the problem to real number division. For example:

$$\frac{(3 + j4)}{(2 + j2)} = \frac{(3 + j4)(2 - j2)}{(2 + j2)(2 - j2)} = \frac{6 + j8 - j6 - j^2 8}{4 + 4}$$

$$= \frac{6 + j2 + 8}{8} = \frac{14 + j2}{8} = 1.75 + 0.25j$$

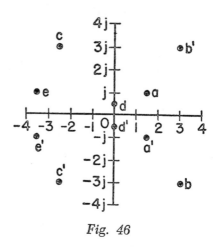

Fig. 46

In the above example, both numerator and denominator were multiplied by the complex conjugate of the denominator. The only "division" that was actually carried out was to divide by 8, a real number, in the last step. In the above answer, $j(\frac{1}{4})$ was written $0.25j$ instead of $j0.25$. Either form is correct.

POLAR FORM OF COMPLEX NUMBERS. Complex numbers can be written in a polar form such as:

$$4.3e^{2j}, \quad 5e^{j}, \quad 3e^{0.5j}, \quad 4e^{-0.75j}$$

As shown in Fig. 47, where these points are plotted, any complex number can be represented in this notation by a term of the form ce^{dj} where c is the distance in the complex plane from the origin to the point and d is the angle in

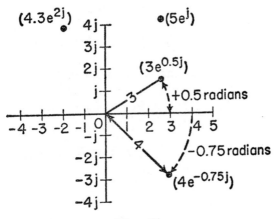

Fig. 47

radians* from the real number base line to a ray from the origin to the point.

Any complex number can be written in either Cartesian form $(a + jb)$ or in polar form (ce^{dj}).

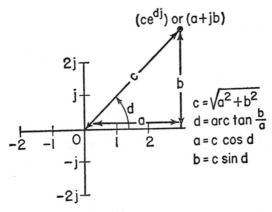

Fig. 48

Figure 48 shows the equivalence of the two forms and shows their conversion equations. The fact that the two

* One radian = 57.3 degrees. 6.28 radians = 2π radians = 360 degrees = one cycle.

forms are equivalent is not arbitrary, for either can be derived from the other by using the basic infinite series definitions for $\sin x$, $\cos x$ and e^x that are found in any handbook.

Notice in Fig. 48 that since the magnitude of a complex number is given by c in ce^{jd}, then e^{jd} by itself must have a magnitude of unity. This is true and a few values of e^{jd} are plotted in Fig. 49.

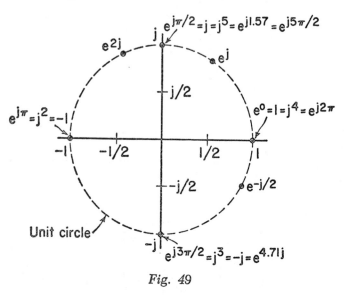

Fig. 49

Notice also that all the points shown in Fig. 49 lie on the "unit circle"—a circle of unit radius that is the dividing line between complex numbers of magnitude less than one and complex numbers of magnitude greater than one.

Since $e^{j2\pi} = 1$, d in e^{jd} can be increased or decreased by any integer number times 2π without changing the value of the complex number. Thus $e^{j\pi/2} = e^{j(\pi/2+2\pi)}$. Since $j = e^{j\pi/2}$, these notations can be interchanged at will. It is suggested that the reader check all the equalities of Fig. 49 as an exercise.

Two complex numbers in polar form can be multiplied together very easily. The magnitude of their product is the product of the individual magnitudes of the multipliers and the angle of the product is the sum of the angles of the multipliers. That is,

$$Ae^{aj} \times Be^{bj} = ABe^{(a+b)j}$$

This is illustrated in Fig. 50.

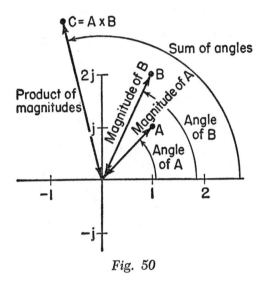

Fig. 50

Division is the inverse of multiplication. Thus to divide complex numbers in polar form, their angles are subtracted and their magnitudes are divided.

POWERS AND ROOTS. Raising a complex number to a power is an extension of multiplication and consists of *multiplying* the angle by the power and simultaneously raising the magnitude to the power. For example:

$$(4e^{j2})^3 = (4)^3 e^{j(2 \times 3)}$$
$$= 64e^{6j}$$

Finding the roots of a complex number is equivalent to raising the number to a fractional power. For example:

$$(4e^{j2})^{0.5} = (4)^{0.5} e^{j(2 \times 0.5)} = 2e^{j}$$

There is a second square root, however, since

$$4e^{j2} = 4e^{j(2+2\pi)} = 4e^{j8.28}$$

Then it is also true that

$$(4e^{j2})^{0.5} = (4e^{j8.28})^{0.5} = (4)^{0.5}e^{j(8.28\times0.5)}$$

$$= 2e^{j4.14}$$

The two square roots of $4e^{j2}$ are shown in Fig. 51.

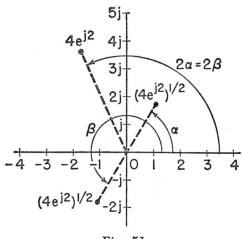

Fig. 51

Notice that the two square roots are 180 degrees apart and that doubling the angle of either square root yields the angle of the original number. By a similar approach, three cube roots, four fourth roots, etc. can be found for any number, real or complex.

Vectors, phasors, symmetrical components, and complex numbers are often confused and misinterpreted as being equivalent. All are added in a similar manner, but each has separate features. Vectors are multiplied in two different fashions (called dot and cross). Phasors are complex numbers that are on a complex plane that is considered to be spinning about the origin at a constant speed with respect to a set of fixed axes. At any fixed instant of time, phasors are complex numbers. Symmetrical components are an extension of phasors such that there are three complex planes in use in all calculations. One complex plane spins clockwise, the second spins counterclockwise, and the third remains fixed. (These three planes are mathematical tools used to reduce non-symmetrical, three-phase electrical problems into three symmetrical problems.)

8. COMPLEX FUNCTIONS

FUNCTIONS OF A COMPLEX VARIABLE. A function of a real variable is defined as:

$$y = f(x)$$

Where y is a function of x. In words, for every value of x, there is a value (or more than one value) of y. Similarly, a function of a complex variable is defined by

$$z = f(w)$$

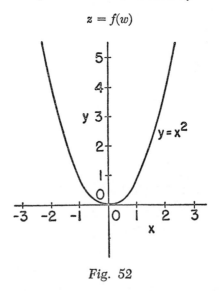

Fig. 52

where, for every value of w there is a value (or more than one value) of z. In this case, w and z are complex numbers. A difference between real and complex number functions becomes apparent in plotting them. If for example,

$$y = x^2$$

and x and y are restricted to be real numbers, any value of x or y can be represented as a point on a line, a single dimension. Two dimensions (a plane) are therefore sufficient to plot y as a function of x as shown in Fig. 52.

In the case of complex numbers, two axes are needed to plot either z or w alone; therefore, two sets of axes are needed to plot a function.

Figure 53 shows the plot of $z = w^2$ where z and w are both complex numbers. In order to read the graph it is necessary to pick a point on the w plane; that is, select a value of w; then find the corresponding point on the z plane and read the desired value of z from the z-plane axes. For example, if w is $1.5 + j0$ (point "e" on the w plane), z is $2.25 + j0$ (point "e" on the z plane). A dashed grid is drawn in both planes to facilitate finding points in either plane that correspond to points in the other plane. It is suggested that the reader find points "a" through "f" in both planes as an exercise.

Notice that points "e" and "f" are the same point in the z plane and that half of the w plane corresponds to the entire z plane. This is because, if $y = x^2$ in real numbers or $z = w^2$ in complex numbers, either x or w has two values for every single value of y or z. In other words, x and w are multiple-valued functions of y and z, respectively. The technique of plotting a grid on the w plane to cause a distorted grid to appear on an $f(w)$ plane is used as a tool to transform problems from one geometry to another and is called transformal mapping.

Another way to plot a function of a complex variable is to plot only the magnitude of the function in the form of a plaster model. This is shown in Fig. 54 for the case of $z = w^2$.

Only a part of the w plane is illustrated and the top of the model has been cut off because its height (the magnitude of z) increases rapidly as the magnitude of w increases.

Of particular interest in design and anaylsis problems are the occurrence of "zeros" and "poles" in the three-dimensional plot of a complex function. A zero is a point on the complex plane of the independent variable (w) where

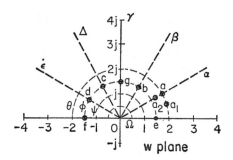

$$z = w^2$$

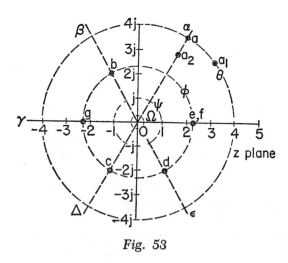

Fig. 53

the dependent variable (z) has a magnitude of zero. Such a point occurs at the origin in Fig. 54 for the case of $z = w^2$.

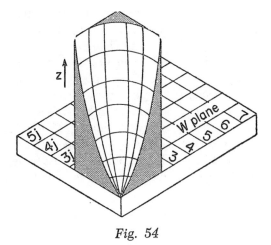

Fig. 54

A pole is one type of singular point. A singular point is a point in the plane of the independent variable where the dependent variable becomes infinite. A singularity occurs in

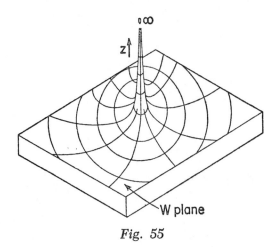

Fig. 55

the plot of z vs w shown in Fig. 55 at the point in the w plane where z becomes infinite.

There are two types of singularities—removable and essential. Loosely speaking, a singularity at which the function goes to infinity with ordinary steepness is a "removable singularity" or "pole." A singularity at which the function goes to infinity with extraordinary steepness is an "essential singularity." The precise test to determine the type of singularity involved is to replot the function after multiplying it by $(w - w_1)$ where w_1 is the complex value of w at which the singularity occurs. If the plot of $f(w) \cdot (w - w_1)$ has no singularity at $w = w_1$, then the singularity is a "pole of first order" (also written "first order pole"). If the singularity remains after the above step but disappears in a plot of $f(w) \cdot (w - w_1)^2$, then it is a pole of second order, and so on. If the pole is of infinite order, it is called an essential singularity.

9. COMPLEX CALCULUS

DIFFERENTIATION OF COMPLEX NUMBERS. The derivative of a function of a complex number is similar to the derivative of a function of a real number. If z is a function of w, dz/dw is the limit of the ratio of a very small change in z to the corresponding small change in w as the size of the change is reduced to zero. As an example, the derivative of z with respect to w, dz/dw, for the case of $z = w^2$, will be found at point a in Fig. 53 by an approximate calculation using a finite change in w.

If z goes from a to a_1 in the z plane, w goes from a to a_1 in the w plane. The desired derivative is then

$$\frac{dz}{dw} = \frac{z_{a1} - z_a}{w_{a1} - w_a} = \frac{(3.1 + 2.5j) - (1.9 + 3.5j)}{(1.8 + 0.6j) - (1.7 + j)}$$

$$= \frac{1.2 - j}{0.1 - 0.4j} = \frac{(1.2 - j)(0.1 + 0.4j)}{(0.1 - 0.4j)(0.1 + 0.4j)}$$

$$= \frac{0.12 + 0.48j - 0.1j + 0.4}{0.01 + 0.16} = \frac{0.52 + 0.38j}{0.17}$$

$$= 3.1 + 2.2j$$

$$= 2(1.65 + 1.1j) = 2w_a \quad \text{(approximately)}$$

Thus the derivative dz/dw evaluated at a is approximately twice the value of the dependent variable, w, at point a. In symbols, $dz/dw = 2w$. The result would have been the same if points a and a_2 had been used instead of points a and a_1. In real number calculus, if $y = x^2$, then $dy/dx = 2x$. Thus it appears that the differentiation formulas of func-

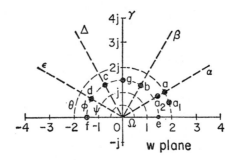

$$\boxed{z = w^2}$$

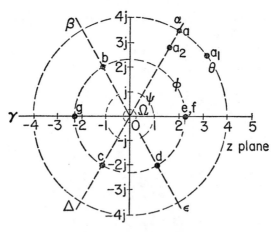

Fig. 53 *(repeated)*

tions of real numbers apply to functions of complex numbers as well. In general they do, but there are numerous exceptions. The differentiation formulas of trigonometric functions (sin x, tan x, etc.) and integer power functions (x^2, x^3, etc.) apply to complex numbers except at singular points.

If the derivative is thought of as a complex number in polar form, it can be visualized rather easily. If z is a function of w, the magnitude of the derivative, dz/dw, is the ratio of a change in z on the z plane to the corresponding change in w on the w plane and the angle of the derivative is the angle between the plot of the change in z on the z plane and the plot of the change in w on the w plane if the two planes were superimposed. This effect can be seen in Fig. 53 by comparing $a - a_2$, $a - a_1$ and $a_1 - a_2$ in the z- and w-plane cases.

Analytic Functions are essentially well-behaved functions. They have no discontinuities or singularities at any point,* and the derivative of the function exists at every point. By "the derivative exists" is meant that the derivative is a finite complex number and is the same value no matter what small change is assumed in the independent variable while calculating the derivative.

A function may be analytic over most of the plane of the independent variable (part of a plane is sometimes called a region) and non-analytic in a few small regions. Most complex functions which arise in the application of complex variables to the solution of physical problems are of this type. As an example and review of terms: the function $z = w/(w - 7j)^2$ has a zero at $w = 0$, a pole of second order at $w = +7j$, and is analytic over the entire w plane except at $w = 0 + 7j$.

INTEGRATION OF COMPLEX FUNCTIONS. The integral of a complex function, say $z = f(w)$, is the sum obtained by multiplying the function (z) by small changes in the independent variable (w) over a chosen path between two limits (say, $w = w_a$ and $w = w_b$). In symbol form such an integral is written:

$$\int_{w=w_a}^{w=w_b} z \, dw$$

Ordinarily a complex integral is evaluated by breaking it up into a sum of two or more integrals such that some compo-

* Within limits to be defined later.

nent of w does not change during the integration, in order to simplify the calculations.

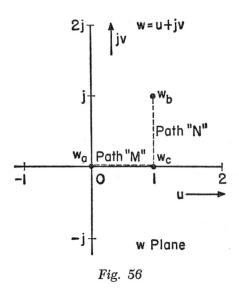

Fig. 56

For example, to evaluate

$$\int_{w_a=0}^{w_b=1+j1} z \, dw$$

where

$$z = w^2$$

a path is chosen as shown in Fig. 56. (In the case of analytic functions it can be shown that the value of the integral is independent of the path of integration.) Along this path:

$$\int_{w_a}^{w_b} z \, dw = \int_{w_a}^{w_c} z \, dw + \int_{w_c}^{w_b} z \, dw$$

$$\text{(path M)} \quad \text{(path N)}$$

For path M, w and z are real numbers. Using the notation that $w = u + jv$, for path M, $z = w^2 = u^2$. Since only the u component of w changes along path M, $dw = du^*$ and

* The change of variables during the evaluation of an integral will be explained further in Chapter 12.

the integral for path M becomes:

$$\int_{w_c}^{w_b} z\, dw = \int_{w=0}^{w=1} w^2\, dw = \int_{u=0}^{u=1} u^2\, du = \left[\frac{u^3}{3}\right]_0^1 = +\frac{1}{3}$$

In the case of path N, $w = 1 + jv$, v varies from 0 to 1 and $dw = d(jv)$ so that the integral for path N becomes:

$$\int_{w_c}^{w_b} z\, dw = \int_{v=0}^{v=1} (1 + jv)^2\, d(jv) = j\int_{v=0}^{v=1} (1 + 2jv - v^2)\, dv*$$

(path N)

$$= j\int_{v=0}^{v=1} dv + 2j^2 \int_{v=0}^{v=1} v\, dv - j\int_{v=0}^{v=1} v^2\, dv$$

$$= \left[jv + \frac{2j^2 v^2}{2} - \frac{jv^2}{3}\right]_0^1$$

$$= j - 1 - \frac{j}{3} = -1 + \frac{2}{3}j$$

Adding the two path integrals (M and N) gives the total integral value as $\frac{1}{3} - 1 + \frac{2}{3}j = -\frac{2}{3} + \frac{2}{3}j$. Notice that since j is a constant number, complex integration requires no more techniques than real integration for the case of analytic functions. In the case of analytic functions the integral around any closed path (complete loop) is zero.

RESIDUES. In the case of a function that is analytic except at a pole, the integral around a closed path that goes around the pole is not zero. It has a value of $2\pi jR$ where R is a number called the "residue" of the pole if the path is traversed in a counterclockwise direction.

If the path of integration encircles several poles, the integral has a value of $2\pi j$ times the sum of the residues of the poles. The residue can be thought of as a measure of the abruptness of a pole.

To find the residue of a pole of order m of a function $f(w)$ of a complex number, (w), where the pole exists at a value of $w = w_0$,

---

* The validity of removing j from the $d(jv)$ term and placing it in front of the integral sign will be explained in Chapter 12. For the present accept it as a valid technique.

A. Write a new function of w called s where

$$s = \frac{(w - w_0)^m f(w)^*}{(m - 1)!}$$

B. The residue, R, is the value of the $(m - 1)$st derivative of s with respect to w at the point $w = w_0$.

$$R = \left[\frac{d^{(m-1)}s}{dw^{(m-1)}}\right]_{w=w_0}$$

As an example,

$$f(w) = \frac{w + 7}{w^2(w - 3)}$$

has a pole of first order at $w = 3$ and a pole of second order at $w = 0$. The residue at the first-order pole, $w = 3$, can be found as follows:

A. $\qquad s = \dfrac{(w - 3)^1(w + 7)}{(1 - 1)!w^2(w - 3)} = \dfrac{w + 7}{0!w^2}$†

$$s = \frac{w + 7}{w^2} = w^{-1} + 7w^{-2}$$

B. $R_{(w=3)} = \left[\dfrac{d^0s}{dw^0}\right]_{w=3} = [s]^{‡}_{w=3} = \dfrac{1}{3} + \dfrac{7}{9} = \dfrac{10}{9}$

The residue at the second-order pole, $w = 0$, can be found similarly:

A. $\qquad s = \dfrac{(w - 0)^2(w + 7)}{1!w^2(w - 3)}$

$$s = \frac{w + 7}{(w - 3)}$$

* $(m - 1)!$ is read "$m - 1$ factorial" and means $(m - 1) \times (m - 2) \times (m - 3)$ etc. down to the one term. For example: $4! = 4 \times 3 \times 2 \times 1 = 24$. Another example: $6! = 6 \times 5 \times 4 \times 3 \times 2 \times 1 = 720$.

† Zero factorial equals one.

‡ The zero order derivative is the function itself.

B. $R_{(w=0)} = \left[\dfrac{ds}{dw}\right]_{w=0} = \left[\dfrac{d}{dw}\left(\dfrac{w+7}{w-3}\right)\right]_{w=0}$

$R_{(w=0)} = \left[\dfrac{-10}{(w-3)^2}\right]_{w=0} = \dfrac{-10}{9}$

Notice that in the case of the simple pole, s is evaluated directly to determine the residue.

Residue evaluation and pole location are key elements in Laplace operational calculus as will be seen in Chapter 11.

V

OPERATIONAL CALCULUS

10. TRANSFORMATIONS

SOLVING EQUATIONS. "Solving" an equation usually means rearranging the terms of the equation until one variable of interest is most concisely expressed in terms of another variable. An example will illustrate the usual algebraic method of solving equations and indicate the need for operational calculus.

Given the problem:

Solve for x as a function of y if

$$y^2x + 2yx + 3 = 6$$

The first step is to collect all terms containing x on one side of the equation, which yields

$$y^2x + 2yx = 3$$

Then, factor x out to obtain

$$x(y^2 + 2y) = 3$$

from which

$$x = \frac{3}{y^2 + 2y}$$

Algebraic quantities were multiplied, divided and factored to reach the solution.

The algebraic processes above could not be used to solve a linear differential equation such as the following:

$$\frac{d^2x}{dt^2} - \frac{3dx}{dt} + 2x = 1$$

since (d^2x/dt^2) does not equal (dx/dt) times (dx/dt) nor does

$$\frac{dx}{dt} = x \frac{d}{dt}$$

TABLE VII

LAPLACE TRANSFORMS

Time Function (x)	Laplace Transform (X)
1	$\dfrac{1}{p}$
unit step at $t = 0^+$	$\dfrac{1}{p}$
e^{-kt}	$\dfrac{1}{p + k}$
t	$\dfrac{1}{p^2}$
t^2	$\dfrac{2}{p^3}$
t^n	$\dfrac{n!}{p^{(n+1)}}$
te^{-kt}	$\dfrac{1}{(p + k)^2}$
$\sin kt$	$\dfrac{k}{p^2 + k^2}$
$\cos kt$	$\dfrac{p}{p^2 + k^2}$
$\dfrac{1}{(n - 1)!} t^{(n-1)} e^{-kt}$	$\dfrac{1}{(p + k)^n}$
$e^{-at}(E \cos kt + F \sin kt)$	$\dfrac{E(p + a) + Fk}{(p + a)^2 + k^2}$

The usual manual procedure is to guess at the form of the answer and solve the differential equation by refined trial and error techniques. If it were possible to substitute an algebraic term for d/dt and solve differential equations by algebraic techniques, the problem would be simplified.

Such a process of substitution is possible and useful and is called Laplace operational calculus.* One example of the Laplace technique was given at the close of Chapter 6. Another example of the use of Laplace operational calculus will be shown now. Each step of the example will be explained in detail later in this chapter or in Chapter 11.

Assume it is desired to solve

$$\frac{d^2x}{dt^2} - \frac{3dx}{dt} + 2x = 1 \tag{82}$$

where $dx/dt = x = 0$ at time $t = 0$.

Using lists of conversions such as shown in Tables VII and VIII,† Equation 82 can be converted to:

$$p^2X - 3pX + 2X = \frac{1}{p} \tag{83}$$

where p‡ is the transform of (stands for) d/dt, X is the transform of x, $1/p$ is the transform of 1, and p and X are algebraic terms.

Solving for X in Equation 83 by algebraic processes gives

$$X = \frac{1}{p(p^2 - 3p + 2)} = \frac{1}{p(p - 1)(p - 2)}$$

or

$$X = \frac{0.5}{p} - \frac{1}{p - 1} + \frac{0.5}{p - 2} \tag{84}$$

Using Table VII to transform the terms in Equation 84 back

---

* Laplace is the most popular operational calculus. Many others are possible.

† Page 128.

‡ S is used in place of p in some texts.

to functions of time gives:

$$x = 0.5 - e^t + 0.5e^{2t}$$

which is the solution of Equation 82.

The Laplace operational calculus process used above consists of the following steps:

1. Transform the differential equation where each term is a function of time, t, into an algebraic equation where each term is a function of p, an algebraic quantity.

2. Manipulate to solve for X, the transform of the unknown term, x.

3. Transform back (inverse transform) to obtain x as a function of time.

THE LAPLACE TRANSFORM. The Laplace Transform, abbreviated $L(\ )$, of any function of time, x, is given by:

$$L(x) = \int_{t=0}^{t=+\infty} e^{-pt}x \, dt \qquad (85)$$

This integral can be interpreted as shown in Fig. 57.

Figure 57A shows a plot of x as a function of time. The solid curve of Fig. 57B shows a typical plot of e^{-pt} as a function of time if p is a real number greater than zero.

The product of x and e^{-pt} is shown in Fig. 57C. The Laplace transform is the area under the curve of (xe^{-pt}) from time $t = 0$ to $t = \infty$ and is shown shaded in Fig. 57C. This area is a finite number and will have a specific number value for a given number value of p. The area when finally calculated is a function of p but not a function of t or x. The function of p that results (i.e., how the area varies with p) depends upon the shape of the x vs t curve, but the Laplace transform is an expression of p alone. (See Table VII.)

Had a different value of p been chosen, the plot of e^{-pt} would have changed, possibly to the dashed curve of Fig. 57B, and the Laplace transform, the area under the curve in Fig. 57C would have been smaller. For a specific number value of p, two different functions of time might have the same area, but only identical time functions will have identical areas for all values of p. The expression e^{-pt}

serves two purposes in the Laplace transform integral (Equation 85). The term e^{-pt} makes the transform exist for functions of time that do not go to zero as time goes to infinity and it guarantees that each time function has a unique transform.

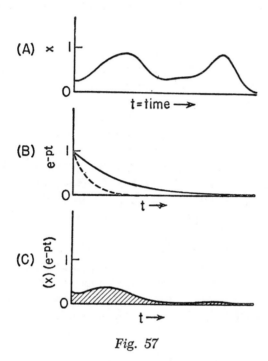

Fig. 57

As an example of the calculation of a Laplace transform, consider

$$L(e^{-t}) = \int_{t=0}^{=\infty} e^{-pt}(e^{-t}) \, dt$$

which can be directly integrated since p is not a function of time and therefore is constant during the integration. Thus:

$$L(e^{-t}) = \int_{t=0}^{t=\infty} e^{-(p+1)t} \, dt$$

$$= \left[\frac{-e^{-(p+1)t}}{p+1} \right]_{t=0}^{t=\infty} = 0 - \frac{-e^{-0}}{p+1} = \frac{1}{p+1}$$

Therefore

$$L(e^{-t}) = \frac{1}{p + 1}$$

So that the Laplace transform of the function $x = e^{-t}$ is $X = 1/(p + 1)$. All the transforms of Table VII can be derived by similar means.

LIMITATIONS OF LAPLACE TRANSFORMS. Most functions of time ordinarily encountered have Laplace transforms. In order to have a Laplace transform, a function must be finite and continuous (except that it may have finite discontinuities) for all values of t greater than zero. In addition, as t goes to infinity, the function must not grow at a faster rate than e^{-pt} decays, for if it does, the area under the curve in Fig. 57C is infinite. The restriction that the function be finite is not a firm restriction. However, if the function becomes infinite at any value of time, it must do so in such fashion that the area under the xe^{-pt} curve remains finite.

Laplace transforms may be added, subtracted, multiplied, etc. like any algebraic quantities; however, combining time functions may affect their transforms in odd ways. Thus, if x and y are time functions and X and Y are their respective Laplace transforms, the Laplace transform of their sum is the sum of their transforms:

$$L(x + y) = L(x) + L(y) = X + Y$$

and if R is any number,

$$L(Ry) = RL(y) = RY$$

but

$$L(x \div y) \text{ does not equal } X \div Y$$

and

$$L(x \cdot y) \text{ does not equal } X \cdot Y$$

However

$$X/Y = X(1/Y)$$

This is not too surprising, for the operation of finding the Laplace transform of a time function is an integration process and integration rules apply. For example:

$$\int (x + y)\, dz = \int x\, dz + \int y\, dz$$

but

$$\int (xy)\, dz \text{ does not equal} \left(\int x\, dz \right)\left(\int y\, dz \right)$$

It should be kept in mind that the Laplace transform of the type described in this book has no meaning for values of time t less than zero.

Although the Laplace transform is a function of p and the original function is a function of t and both are linked by an equation, there is no direct correspondence between t and p. As a result, it cannot be said that a particular value of t corresponds to a particular value of p. There are a few cases where particular values of the function of time can be directly evaluated by putting number values in for p and they will be discussed later, but they are the exception; not the rule.

TRANSFORMS OF DERIVATIVES AND INTEGRALS. It can be derived (by ordinary integration) that the Laplace transform of a function and the Laplace transform of the first time derivative of a function differ by a factor of $1/p$, so that if

$$L(x) = X$$

then

$$L(dx/dt) = pX - (x)_{0+}$$

and

$$L(d^2x/dt^2) = p^2X - p(x)_{0+} - (dx/dt)_{0+}$$

The general form is shown in Table VIII. Also listed in Table VIII are other characteristics of Laplace transforms which are discussed later in this chapter.

Notice that the initial values of x and its derivatives enter into determining the Laplace transform. To interpret

TABLE VIII

Time Function	Laplace Transform	Notes
x	X	notation
$\dfrac{dx}{dt}$	$pX - (x)_{0^+}$*	first derivative
$\dfrac{d^2x}{dt^2}$	$p^2X - p(x)_{0^+} - (dx/dt)_{0^+}$	second derivative
$\dfrac{d^nx}{dt^n}$	$\begin{bmatrix} p^nX - p^{(n-1)}(x)_{0^+} \\ - p^{(n-2)}(dx/dt)_{0^+} \\ - p^{(n-3)}(d^2x/dt^2)_{0^+} \\ - \cdots - \\ (d^{(n-1)}(x)/dt^{(n-1)})_{0^+} \end{bmatrix}$	general derivative
$\displaystyle\int_{-\infty}^{t} x\,dt$	$\dfrac{X + \left[\displaystyle\int_{-\infty}^{0} x\,dt \right]}{p}$	integral
$\displaystyle\int_{-\infty}^{t} x\,dt$	$\dfrac{X}{p}$	integral with zero constant of integration
$\displaystyle\int_{0}^{t} \left[\int_{0}^{t} x\,dt \right] dt$	$\dfrac{X}{p^2}$	repeat of above
$e^{-kt}x$	$X(p + k)$	shifting theorem
$x(t - t_1)$	Xe^{-t_1p}	time delay†
$x + y$	$X + Y$	addition of transforms
kx	kX	multiplication by a constant
$\lim_{t\to\infty} (x)$	$= \lim_{p\to0} (pX)$	final value†
$\lim_{t\to0} (x)$	$= \lim_{p\to\infty} (pX)$	initial value

* 0^+ means "at $t = 0^+$."

† See text for restrictions.

the notation used, $(x)_{0^+}$ means "the numerical value of x at time $t = 0^+$" and $(dx/dt)_{0^+}$ means "the numerical value of dx/dt at time $t = 0^+$."

A similar relationship using factors of $1/p$ holds for integration with respect to time. For example:

$$L\left(\int_{t=-\infty}^{t=t} x\, dt\right) = \frac{1}{p}\left(X + \int_{t=-\infty}^{t=0} x\, dt\right)$$

The second term in the parentheses of the transform corresponds to the constant of integration and is the numerical value of the integral at time $t = 0^+$. For the case where the value of the integral is zero at time $t = 0^+$, the initial condition term is zero, leaving

$$L\left(\int_{t=0}^{t=t} x\, dt\right) = \frac{X}{p}$$

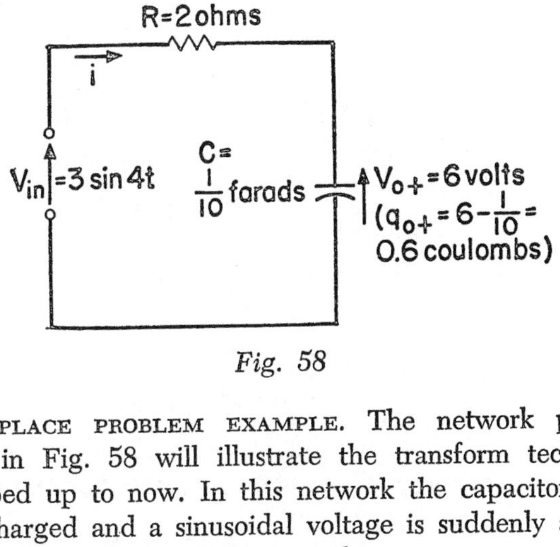

Fig. 58

LAPLACE PROBLEM EXAMPLE. The network problem shown in Fig. 58 will illustrate the transform techniques developed up to now. In this network the capacitor is initially charged and a sinusoidal voltage is suddenly applied.

From basic electrical circuit theory:

$$V_{\text{in}} = iR + \frac{1}{C}\int i\, dt \tag{86}$$

Taking the Laplace transform of each side of Equation 86 yields

$$L(V_{\text{in}}) = RL(i) + \frac{1}{C}L\left(\int i\, dt\right) \tag{87}$$

The first term of Equation 87 is evaluated by using Table VII.

$$L(V_{in}) = L(3 \sin 4t) = 3L(\sin 4t) = \frac{12}{p^2 + 16} \qquad (88)$$

The transformation of the second term of Equation 87 is:

$$RL(i) = RI = 2I \qquad (89)$$

where $I =$ the Laplace transform of i. The last term of Equation 87 is an integral so that:

$$\frac{1}{C} L\left(\int i \, dt \right) = \frac{1}{C} \left[\frac{I}{p} + \frac{\int_{-\infty}^{0} i \, dt}{p} \right]$$

or

$$\frac{1}{C} L\left(\int i \, dt \right) = 10 \left(\frac{I}{p} + \frac{q_{0^+}}{p} \right) = 10 \left(\frac{I}{p} + \frac{6}{p} \right) = \frac{10I + 6}{p} \qquad (90)$$

Notice that the initial charge on the capacitor, q_{0^+}, is the time integral of the current from $t = -\infty$ to $t = 0$. Substituting Equations 88, 89, and 90 into Equation 87 gives:

$$\frac{12}{p^2 + 16} = 2I + \frac{10I + 6}{p}$$

or

$$I = \frac{12p - 6p^2 - 96}{(2p + 10)(p^2 + 16)} \qquad (91)$$

Equation 91 can be reduced to the sum of two fractions called "partial fractions." Assuming for the moment that

$$I = \frac{A}{2p + 10} + \frac{Bp + C}{p^2 + 16} \qquad (92)$$

is the partial fraction form desired (this will be discussed further later), the fractions of Equation 92 can be reduced to a single common denominator fraction to give

$$I = \frac{A(p^2 + 16) + (Bp + C)(2p + 10)}{(2p + 10)(p^2 + 16)}$$

$$= \frac{Ap^2 + 16A + 2Bp^2 + 10Bp + 2pC + 10C}{(2p + 10)(p^2 + 16)}$$

$$= \frac{(A + 2B)p^2 + (10B + 2C)p + (16A + 10C)}{(2p + 10)(p^2 + 16)} \quad (93)$$

By comparing Equation 93 with Equation 91, it is apparent that

$$-6 = A + 2B$$

$$12 = 10B + 2C$$

$$-96 = 16A + 10C$$

These can be solved to give (approximately)

$$A = -7.46$$

$$B = 0.732$$

$$C = 2.34$$

Substituting these values into Equation 92 leads to:

$$I = \frac{-7.46}{(2p + 10)} + \frac{0.732p}{(p^2 + 16)} + \frac{2.34}{(p^2 + 16)} \quad (94)$$

Using Table VII, this Laplace transform can be inverse transformed to the following time function:

$$i = -3.73e^{-5t} + 0.732 \cos 4t + 0.585 \sin 4t \quad (95)$$

which is plotted in Fig. 59.

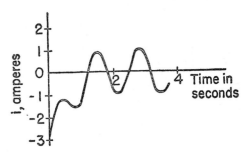

Fig. 59

It is suggested that the reader check the transition from Equation 94 to Equation 95 as an exercise. Notice that Equation 95 contains both transient and steady-state terms. Also, except for the partial fraction derivation, there was no need for trial and error or guessing at the form of the answer. Laplace operational calculus yields fairly direct answers to most problems.

The sequence of

1. writing the differential equation,
2. transforming to functions of p,
3. manipulating the transformed expressions to solve for the desired variable, and
4. inverse transforming to obtain the answer in normal form is the typical pattern for Laplace operational calculus solutions of differential equations.

11. INVERSE TRANSFORMATIONS

THE INVERSE TRANSFORM. The inverse transform can be evaluated by an integration process using complex numbers. Up to now, p has been regarded as a real number, but everything said so far holds as well if p is a complex number as long as its real part is positive. (If the real part of p is positive, then e^{-pt} in the Laplace transform integral, Equation 85, causes the area under the curve in Fig. 57C to be finite.)

If p is thought of as a complex number, then the Laplace transform of a time function is a function of a complex number and has poles and zeros at various values of p. The location and residues of the poles of Laplace transforms are key elements in inverse transformation. In evaluating the inverse transform, p must be regarded as complex.

Inverse transformation is the process of going from the p function to the corresponding t function. The inverse transform is an integration process and can be shown to be equivalent to

$$x = L^{-1}(X) = \frac{1}{2\pi j} \int_{p=d-j\infty}^{p=d+j\infty} e^{pt} X \, dp \qquad (96)$$

where the path of integration is a straight line. This expression will now be explained by reference to Fig. 60.

Equation 96 states that x, the inverse transform of X, equals a constant $(1/2\pi j)$ times a line integral in the p plane of e^{pt} times X.

Notice that in this integration t is a constant just as p was a constant in the integral used to derive the Laplace transform (X) of a time function (x). Another point of interest is that the term e^{pt} has no singularities or poles except as the real part of p goes to $+\infty$ off to the right side of Fig. 60. As a result, the location of any poles of $e^{pt}X$

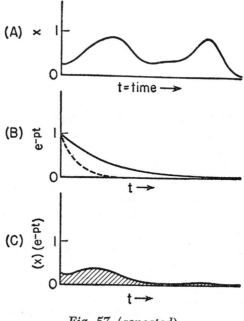

Fig. 57 (repeated)

are at the location of poles of X. One such pole is indicated in Fig. 60.

An assumption that will be needed in this discussion is that

$$\int X e^{pt}\, dp = 0 \qquad (97)$$

around the dot-dash semicircle at infinity shown in Fig. 60. This is true for all functions of time that are powers of t, for $\sin kt$, $\cos kt$, e^{kt} and many other functions. A safe rule to follow is that Equation 97 is true for any function of p which, when written as a fraction in unfactored form, contains a higher power of p in the denominator than in

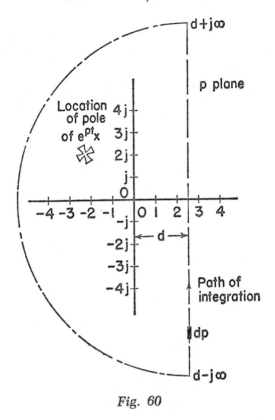

Fig. 60

any terms in the numerator. A look at Table VII shows that all the transformed expressions in the table meet this criterion.

The distance d is put in Equation 96 and the diagram of Fig. 60 primarily to assure that the path of integration lies to the right of any poles of X.

By the theory of residues presented in Chapter 9, the integral of $e^{pt}X$ around the dashed line-dot-dashed line closed path in Fig. 60 is equal to

$$\int e^{pt}X \, dp = 2\pi j R_1 \tag{98}$$

where R_1 is the residue of the pole of $(e^{pt}X)$, assuming that $(e^{pt}X)$ has only one pole in the region to the left of $p = d$ as shown in Fig. 60.

The integral of Equation 98 has the same value as the line integral around a closed path just large enough to include the pole (as discussed in Chapter 9). In the small closed path case, however, if the pole is a first-order pole, e^{pt} can be considered as a constant over the small region described at a value of $e^{p_1 t}$ where p_1 is the location of the pole; so in the first-order pole case, e^{pt} can be removed from under the integral sign. Therefore,

$$\int e^{pt} X \, dp \;=\; 2\pi j R_1 \;=\; e^{p_1 t} \int X \, dp \;=\; 2\pi j e^{p_1 t} R_x \quad (99)$$

(closed path)   (small closed
                      path around p_1)

where R_x is the residue of X at the given pole. Combining Equation 99 with Equation 97 gives:

$$x = L^{-1}(X) = \frac{1}{2\pi j}(2\pi j e^{p_1 t} R_x) - \int e^{pt} X \, dp$$

(dash-dot line)

but the integral around the dash-dot line is zero, leaving:

$$x = L^{-1}(X) = e^{p_1 t} R_x$$

for the case of first-order poles of X.

In general, for poles of X of any order, to evaluate the inverse transform of X it is necessary only to locate and evaluate the residue of each pole of $X e^{pt}$ on the part of the p plane that lies to the left of a large finite value of p.

It can be shown that for functions with only first-order poles at p_1, p_2, p_3, $\cdots$,

$$x = L^{-1}(X) = R_1 e^{p_1 t} + R_2 e^{p_2 t} + R_3 e^{p_3 t} + \cdots \quad (100)$$

where R_1, R_2, R_3, etc. are the residues of all the poles of X in the p plane to the left of a large finite value of p.

As an example of residue calculations, in the sample problem worked before, there was a term $-3.73/(p + 5)$. This term has a first-order pole at $p = -5$ with a residue of -3.73 so that the inverse transform of the term is $-3.73e^{-5t}$ as derived before. (It is suggested that the reader review at this point the section in Chapter 9 on the calculation of residues.)

POLE LOCATION. Of particular interest in most applications of Laplace operational calculus is the correspondence between the location of poles of X and the form of x.

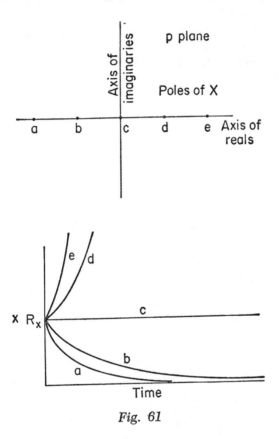

Fig. 61

Figure 61 shows the effect of the location of poles of X on the nature of the time response, x, for the case of first-order poles. A first-order pole on the real axis corresponds to an exponential time function whose time constant* is equal to the reciprocal of the distance from the pole to

* See note at the end of this chapter if uncertain of the meaning of "time constant."

the origin in the p plane. Thus the time responses that are represented by a and e in Fig. 61 have shorter time constants than those represented by b and d. Notice also that poles to the left of the imaginary axis correspond to functions that decay with time and poles to the right of the axis correspond to functions that increase with time. This is a very important basic concept.

A first-order pole of X will never occur off the real axis in the p plane if the corresponding x equation has only real numbers, except for the case where two first-order poles occur in complex-conjugate pairs.

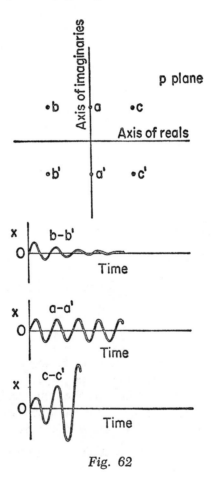

Fig. 62

Figure 62 shows the time function that corresponds to complex-conjugate, first-order pole pairs. Notice that pairs of first-order poles in p correspond to sinusoids of time that decay if the poles are to the left and increase if the poles are to the right of the axis of imaginaries. As an example, $x = \sin 3t$, when transformed, becomes:

$$X = \frac{3}{p^2 + 9}$$

or

$$X = \frac{3}{(p + j3)(p - j3)}$$

which has first-order poles at $p = -j3$ and $p = +j3$.

Figure 63 shows the effect of pole location on the envelope of the corresponding time function for complex conjugate pole pairs to the left of the axis of imaginaries. Notice also that the distance from the axis of reals to either pole of a pair of complex-conjugate first-order poles is the frequency of the sinusoid in radians per second.

If a pole is of second order, it transforms to the shape of the time functions shown in Fig. 61 or 62, except multiplied by t; if it is a third-order pole, by t^2, etc. The cases of a few second-order poles are indicated in Fig. 64 as a guide.

In the case of second-order poles of p on the axis but to the left of the axis of imaginaries, the time to reach peak value in the time function in seconds approximately equals the distance that the point lies to the left of the origin in the p plane, indicated by r in Fig. 64.

As an example of the calculation of an inverse transform by pole location and residues, if a Laplace transform, X, has three poles of first order such that:

	Location	Residue
Pole 1	$p = -3$	$+5$
Pole 2	$p = +2j$	$4j$
Pole 3	$p = -2j$	$-4j$

The corresponding time function is

$$x = 5e^{-3t} + 4je^{2jt} - 4je^{-2jt} \tag{101}$$

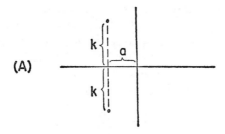

(A)

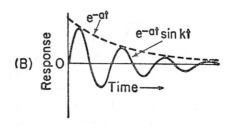

(B)

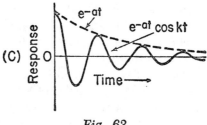

(C)

Fig. 63

where the terms correspond to the poles in the order given. Using the complex number identities that

$$e^{jk} = \cos k + j \sin k$$

$$e^{-jk} = \cos k - j \sin k$$

which can be derived from the relationship between Cartesian and polar complex notations, Equation 101 becomes:

$$x = 5e^{-3t} + 4j(\cos 2t + j \sin 2t) - 4j(\cos 2t - j \sin 2t)$$

or

$$x = 5e^{-3t} + 4j(2j \sin 2t)$$

Therefore,

$$x = 5e^{-3t} - 8 \sin 2t$$

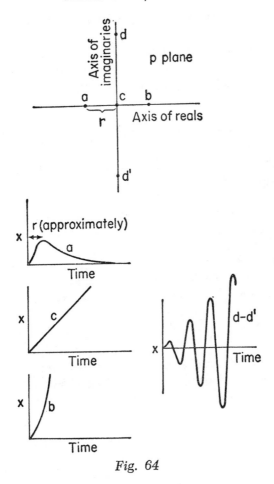

Fig. 64

Notice that poles 2 and 3 formed a complex-conjugate pair and resulted in a sinusoid. Also note that the residue of a pole may be complex.

As a final remark, it is interesting to note that the order, residue, and location of the poles of X completely determine x. This is perhaps to be expected since although x is adequately represented by a single-valued function of time, X is a complex function of p.

PARTIAL FRACTIONS. In order to find the inverse transform of X it is often convenient to reduce X to a sum of many simple fractions called partial fractions whose transforms are known since the inverse transform of a sum of terms is the sum of the individual inverse transforms of the terms. In most cases where partial fractions are useful, X can be expressed as a quotient of two polynomials in p such as

$$X = \frac{mp^4 + np^3 + qp^2 + rp + v}{wp^5 + yp^4 + zp^3 + gp^2 + hp + k} \tag{102}$$

where the highest power of p that appears in the numerator (4 in this case) is lower than the highest power of p that appears in the denominator (5 in this case). If this is not the case, divide the numerator by the denominator until the remainder consists of terms whose denominators are of higher power than their numerators.

The first step in the partial fraction technique is to factor the denominator into linear functions of p and quadratic functions of p. This leads to a form such as:

$$X = \frac{mp^4 + np^3 + qp^2 + rp + v}{(p + a)(p + b)(p + c)^3(p^2 + 2ep + f)} \tag{103}$$

The poles of X occur at $p = -a$, $p = -b$, $p = -c$, and at the roots* of the quadratic factor. The object of the partial fraction expansion is to rewrite X as

$$X = \frac{A}{(p + a)} + \frac{B}{(p + b)} + \frac{C}{(p + c)^3} + \frac{D}{(p + c)^2}$$
$$+ \frac{E}{(p + c)} + \frac{Fp + G}{(p^2 + 2ep + f)} \tag{104}$$

and evaluate A, B, C, D, E, F, and G. Once this is done the inverse transform of each term is easily determined by the method of residues previously described.

The rules to be followed in going from the form of Equation 103 to the form of Equation 104 are as follows:

1. Each linear denominator term such as $(p + a)$ gives rise to a single partial fraction.

* Roots are values of p which make the quadratic term equal to zero.

2. Each repeated linear denominator term such as $(p + c)^3$ gives rise to a partial fraction for its original power and all lesser powers.

3. A quadratic denominator term such as $(p^2 + 2ep + f)$ gives rise to a partial fraction with a numerator of the form $(Fp + G)$.

Once the partial fraction is put in the form shown in Equation 104, the constants $(A, B, \cdots G)$ are evaluated by multiplying both sides by the denominator and solving for the constants algebraically. As an example, if

$$X = \frac{12p^3 + 76p^2 + 119p + 27}{(p + 3)^2(p + 1)p} \tag{105}$$

the partial fraction expansion will have the form:

$$X = \frac{A}{(p + 3)^2} + \frac{B}{(p + 3)} + \frac{C}{(p + 1)} + \frac{D}{p} \tag{106}$$

by the rules stated above. (Notice that in each partial fraction the denominator is of higher order in p than the numerator.) Multiplying Equations 105 and 106 by $(p + 3)^2 \cdot (p + 1) \cdot p$ and equating them yields,

$$12p^3 + 76p^2 + 119p + 27$$
$$= Ap(p + 1) + Bp(p + 3)(p + 1) + Cp(p + 3)^2$$
$$+ D(p + 3)^2(p + 1)$$

or

$$12p^3 + 76p^2 + 119p + 27$$
$$= A(p^2 + p) + B(p^3 + 4p^2 + 3p) + C(p^3 + 6p^2 + 9p)$$
$$+ D(p^3 + 7p^2 + 15p + 9)$$

Setting coefficients of equal powers of p equal to each other gives:

$$12 = B + C + D$$
$$76 = A + 4B + 6C + 7D$$
$$119 = A + 3B + 9C + 15D$$
$$27 = 9D$$

Solving these equations by ordinary algebra yields:

$$A = 5$$

$$B = 2$$

$$C = 7$$

$$D = 3$$

Therefore, by inserting these values into Equation 106,

$$X = \frac{5}{(p+3)^2} + \frac{2}{(p+3)} + \frac{7}{(p+1)} + \frac{3}{p}$$

As an exercise it is suggested that the reader locate the poles of each term, evaluate their order and residue for practice and determine x in this case, using Table VII to get:

$$X = 5te^{-3t} + 2e^{-3t} + 7e^{-t} + 3$$

SPECIAL PROPERTIES OF LAPLACE TRANSFORMS. There are several special properties of the Laplace transform that are of use in various design short-cuts. The first is called the "Final Value Theorem" and states that if X is the Laplace transform of x, then the numerical value of x at time $t = \infty$ is given by the value of pX if p is shrunk to zero. In symbol form:

$$\underset{(\text{at } t = \infty)}{(x)} = \underset{(p \to 0)}{\lim} [pX]$$

This equation holds, however, only if there are no poles of pX on the axis of imaginaries or in the right half of the p plane, for such pole locations indicate sinusoidal or exponentially increasing terms in x that do not approach finite fixed values at time $t = \infty$.

A complementary theorem to the "Final Value Theorem" is the "Initial Value Theorem" which states that if X is the Laplace transform of x, then the numerical value of x at time $t = 0$ is given by the limiting value of pX as p is expanded to an infinite value. In mathematical notation:

$$\underset{(\text{at } t = 0)}{(x)} = \underset{(p \to \infty)}{\lim} [pX]$$

As an example, the initial and final values of the previous problem will be calculated. In that problem:

$$X = \frac{12p^3 + 76p^2 + 119p + 27}{(p+3)^2(p+1)p} \tag{105}$$

or

$$X = \frac{12p^3 + 76p^2 + 119p + 27}{p^4 + 7p^3 + 15p^2 + 9p}$$

The value of pX is therefore

$$pX = \frac{12p^3 + 76p^2 + 119p + 27}{p^3 + 7p^2 + 15p + 9} \tag{107}$$

By inspection of Equation 105, pX will have no poles that will invalidate the Final Value Theorem. The final value of x is given by letting p approach zero as a limit. When this is done to Equation 107, pX becomes $27/9 = 3$, which is the final value of x.

The initial value of x is most easily seen by first dividing numerator and denominator of pX (Equation 107) by p^3 to give:

$$pX = \frac{12 + 76/p + 119/p^2 + 27/p^3}{1 + 7/p + 15/p^2 + 9/p^3}$$

As p becomes infinite the value of pX becomes $12/1 = 12$, which is the initial value of x. As a check, the previously derived expression for x:

$$x = 5te^{-3t} + 2e^{-3t} + 7e^{-t} + 3$$

yields the same final and initial values for x by substituting $t = 0$ and $t = \infty$.

A time delay of t_1 seconds has a Laplace transform of e^{-pt_1}. Another way of expressing this is to say that multiplication of the Laplace transform of a function by e^{-pt_1} causes the corresponding time function to be shifted to the right by t_1 seconds. This is illustrated in Fig. 65 for several functions.

Notice that since the original time function is always

considered undefined or zero for negative time, the shifted function is zero for all time less than t_1.

A substitution of $(p + a)$ for p wherever it occurs in the Laplace transform, X, of a function of time, x, is equivalent to multiplying x by e^{-at}, a decaying exponential. This

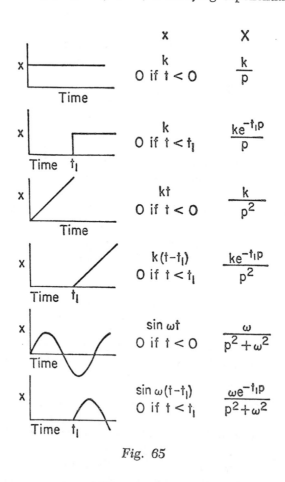

Fig. 65

is called the "Shifting Theorem" and can be interpreted by reference to Fig. 61. If $(p + a)$ is substituted for p, all poles of X shift to the left in the p plane—the direction of more rapid decay of x with time. Substitution of $(p + a)$ for p can lead to errors unless care is taken to notice, for example, that this means $(p + a)^3$ replaces p^3 and *not* that $(p + a)p^2$ replaces p^3.

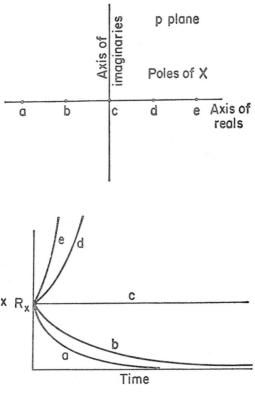

Fig. 61 (repeated)

TIME CONSTANTS. It is very convenient to be able to draw exponentials rapidly when pole and residue techniques are used for inverse transformations. The time constant defines the steepness at $t = 0$ of exponential terms such as $e^{-t/\tau}$ and $e^{t/\tau}$. In both the exponential expressions stated above, τ is the time constant.

Figure 66 shows both exponential curves plotted in terms of their time constants and can be used as a guide for rapidly drawing exponential curves.

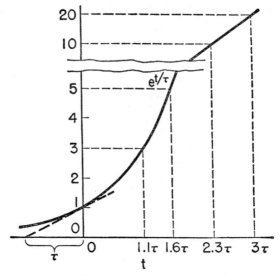

All values are approximate

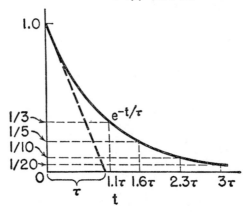

Fig. 66

VI

ADVANCED TOPICS

12. INTEGRATION AND DIFFERENTIATION TECHNIQUES

The solution of calculus problems, whether differential equations or isolated integrals and derivatives, is aided by a knowledge of the techniques discussed in this chapter. It must be kept in mind, however, that any technique should only be used in solving a particular problem if it reduces the total effort needed to solve the problem.

CHANGES OF VARIABLE. It is often desirable both in solving differential equations and in evaluating integrals to change the independent variable. Such changes are simple to perform and are based on differentiation. As an example, if

$$x = v^2 - 9 \tag{108}$$

then

$$\frac{dx}{dv} = 2v \tag{109}$$

so that

$$dx = 2v\,dv \tag{110}$$

Equation 109 expressed the ratio of dx, a small change in x, to dv, the small change in v that corresponds to the small change in x. Equation 110 was derived from Equation 109 by multiplying both sides of Equation 109 by dv. Equation 110 equates two infinitesimals, dx and dv, and so is a different equation from Equations 108 and 109. All three equations, however, express the relationship between x and v. The proper name for infinitesimals like dx and dv is *differentials*.

An example of the application of a change in variables can be seen by using Equations 108 and 110 to simplify this integration problem:

$$\text{Evaluate:} \qquad \int_{x=0}^{x=1} \left(\frac{1}{\sqrt{x+9}} \right) dx \qquad\qquad (111)$$

By substituting (v^2) for $(x + 9)$ and $(2v\, dv)$ for dx *wherever they occur* in Equation 111, thus following the rules set up in Equations 108 and 110, Equation 111 becomes

$$\int_{x=0}^{x=1} \left(\frac{1}{\sqrt{x+9}} \right) dx = \int_{v=3}^{v=3.16} \left(\frac{1}{\sqrt{v^2}} \right) (2v\, dv)$$

$$= \int_{v=3}^{v=3.16} \left(\frac{1}{v} \right) (2v)\, dv$$

$$= \int_{v=3}^{v=3.16} 2dv$$

$$= [2v]_{v=3}^{v=3.16} = 2(3.16 - 3.00) = 0.32$$

In the above solution it should be noticed that the transposition from x to v was carried out in three places: *the function to be integrated, the independent variable of integration,* and *the limits of integration.* A pitfall in the use of changes of variables is that if the variables involved in the change are not single valued monatonic functions of each other, the change in variables may cause gross errors. In the example above, v is a multivalued function of x (for example, at $x = 1$, $v = +3.16$ or -3.16) and the integral was evaluated by consistently using only the positive values of v. In solving a typical problem the relationship between the variables involved in the change (Equation 108) is selected after an examination of the problem to be solved (Equation 111).

The principle of changing variables can be used to develop a useful rule for moving constants around in calculus problems. If a is any constant then

$$\frac{d(ax)}{dx} = a \quad \text{or} \quad d(ax) = a\, dx \qquad\qquad (112)$$

so that multiplying a differential by a constant is equivalent to containing the constant within the differential. An application of Equation 112 can be seen in the evaluation of this integral in which c is a constant:

$$\int_{x=2}^{x=3} e^{cx}\, dx = \frac{c}{c} \int_{x=2}^{x=3} e^{cx}\, dx = \frac{1}{c} \int_{cx=2c}^{cx=3c} e^{cx}\, d(cx)$$

$$= \frac{1}{c} [e^{cx}]_{cx=2c}^{cx=3c} = \frac{1}{c}(e^{3c} - e^{2c})$$

Notice that in the second step a factor of c was moved within the dx differential. An example of this transition in which a constant is removed from within the variable of integration term occurred in the path "N" integration (Fig. 56) in Chapter 8.

RELATIONSHIPS AMONG DERIVATIVES. Reciprocals of derivatives are also derivatives and have meaning.

$$\frac{dy}{dx} = \frac{1}{\dfrac{dx}{dy}} \tag{113}$$

expresses the reciprocal relationship and leads to the obvious conclusion that if $dy/dx = 3$, then $dx/dy = \frac{1}{3}$.

In relationships among many variables it may be cumbersome to compute dz/dx, for example, although dz/dy and dy/dx are simple calculations. A relationship of use in such cases is the chain rule:

$$\frac{dz}{dx} = \frac{dz}{dy} \cdot \frac{dy}{dx} \tag{114}$$

Here again, differentials can be manipulated like any algebraic quantity.

INTEGRATION BY PARTS. Perhaps the most powerful tool for use in integration can be derived from the previously stated equation for the derivative of a product:

$$\frac{d(uv)}{dx} = u\frac{dv}{dx} + v\frac{du}{dx} \tag{115}$$

By integrating both sides of this equation with respect to x (integrals of equals form an equality) and by treating differentials as algebraic quantities,

$$\int \frac{d(uv)}{dx}\, dx = \int u \frac{dv}{dx}\, dx + \int v \frac{du}{dx}\, dx$$

or

$$\int d(uv) = \int u\, dv + \int v\, du$$

from which

$$uv = \int u\, dv + \int v\, du$$

and

$$\int u\, dv = uv - \int v\, du \tag{116}$$

Equation 116 is the "integration by parts" formula and is used to derive analytic expressions. As an example, to evaluate

$$\int x \sin x\, dx = ? \tag{117}$$

the substitutions can be made that

$$u = x \quad \text{and} \quad dv = \sin x\, dx \tag{118}$$

therefore

$$du = dx \tag{119}$$

and by recreating the derivative,

$$\frac{dv}{dx} = \sin x$$

so that

$$v = -\cos x \tag{120}$$

Substituting the equalities of Equations 118, 119, and 120 into the form of Equation 116,

$$\int x \sin x\, dx = \int u\, dv = uv - \int v\, du$$

$$= x(-\cos x) - \int (-\cos x)\, dx$$

so that

$$\int x \sin x \, dx = -x \cos x + \sin x$$

which is the solution of Equation 117. In using the integration by parts formula, the selection of as much of the original integral as "dv" as can be readily integrated is the best key to success. Sometimes the formula leads to an $\int v \, du$ term which must in turn be evaluated by a further application of the integration by parts formula.

NUMERICAL EVALUATION OF DERIVATIVES. If a function is known by its coordinates at discrete points rather than in the form of an analytic expression, its derivatives can be approximately calculated. Such a function is shown in Fig. 67.

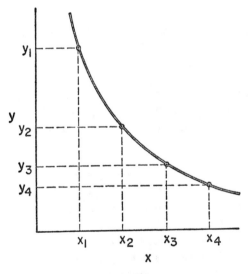

Fig. 67

The first derivative at x_2 can be approximately calculated by dividing the change in y from x_1 to x_2 by the change in x from x_1 to x_2 or

$$\left[\frac{dy}{dx}\right]_{x_1} = \frac{y_2 - y_1}{x_2 - x_1} \tag{121}$$

Similarly, the value of dy/dx at x_2 is approximately

$$\left[\frac{dy}{dx}\right]_{x_2} = \frac{y_3 - y_2}{x_3 - x_2} \qquad (122)$$

The change in dy/dx from x_1 to x_2, when divided by the change in x from x_1 to x_2, yields an approximation of d^2y/dx^2 at x_1. Combining Equations 121 and 122 yields:

$$\left[\frac{d^2y}{dx^2}\right]_{x_1} = \frac{\dfrac{y_3 - y_2}{x_3 - x_2} - \dfrac{y_2 - y_1}{x_2 - x_1}}{x_2 - x_1} \qquad (123)$$

In the case that x_1, x_2, and x_3 are evenly spaced by Δx units, Equation 123 becomes

$$\frac{d^2y}{dx^2} = \frac{(y_1 + y_3) - 2y_2}{(\Delta x)^2}$$

These numerical methods of calculating derivatives can be extended to higher derivatives and are helpful in the use of digital computers to solve differential equations.

EVALUATING ZERO/ZERO. In calculating boundary conditions as part of the process of setting up a differential equation, it is sometimes necessary to evaluate $0 \div 0$. For example,

$$\lim_{(x \to 3)} \left[\frac{x^2 - 9}{x - 3}\right] = ? \qquad (124)$$

In Equation 124 the numerator and denominator both become zero at $x = 3$.

Figure 68 shows a plot of the numerator and denominator of Equation 124 as a function of x. Notice that near $x = 3$ both curves are approximately straight lines and therefore the ratio of the ordinates of the curves in the vicinity of $x = 3$ is the ratio of the slope of the curves. Similarly *if a function is the quotient of two sub-functions which become zero at a limit, the value of the function at the limit is the ratio of the derivatives of the sub-functions.* (This statement assumes that the derivatives of the sub-functions are finite.) As an example of the above principle (which is called L'Hôpital's Theorem), Equation 124 can be evaluated by taking the first derivatives of the numerator and denominator and evaluating their ratio. Thus,

$$\lim_{(x \to 3)} \left[\frac{x^2 - 9}{x - 3} \right] = \operatorname*{Value}_{(x=3)} \left[\frac{2x - 0}{1} \right] = 6$$

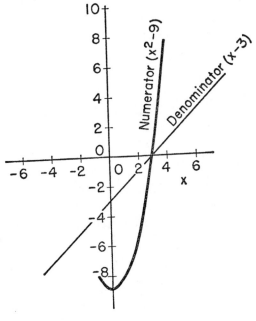

Fig. 68

PARTIAL DERIVATIVES. Sometimes a dependent variable is expressed as a function of more than one independent variable, for example:

Case A: $y = x + 2z$

Case B: $y = x^2z$

where x and z are both independent variables. What would the first derivative or slope of y be? Since y will vary if either x or z varies, a new type of derivative can be defined to describe the partial change of y if one of the independent variables changes at a time. *Therefore, the first partial derivative of y with respect to x is the first derivative of y with respect to x if all other independent variables are assumed to stay constant.* The symbol for this is $\partial y/\partial x =$ the first partial derivative of y with respect to x and in the two cases shown:

Case A: $\quad \dfrac{\partial y}{\partial x} = \left[\dfrac{d(x + 2z)}{dx} \right]_{z \text{ constant}} = 1 + 0 = 1$

Case B: $\quad \dfrac{\partial y}{\partial x} = \left[\dfrac{d(x^2z)}{dx} \right]_{z \text{ constant}} = 2zx$

Similarly, one could deduce $\partial y/\partial z$:

Case A: $\quad \dfrac{\partial y}{\partial z} = \left[\dfrac{d(x + 2z)}{dz} \right]_{x \text{ constant}} = 0 + 2 = 2$

Case B: $\quad \dfrac{\partial y}{\partial z} = \left[\dfrac{d(x^2z)}{dz} \right]_{x \text{ constant}} = x^2$

TOTAL DERIVATIVE. In the event that both independent variables (x and z) change, the resultant change in a dependent variable (y) can be described by what is called the *total derivative* of y.

Total derivative of $y = \dfrac{dy}{d(\ \ )} = \dfrac{\partial y}{\partial x} \cdot \dfrac{dx}{d(\ \ )} + \dfrac{\partial y}{\partial z} \cdot \dfrac{dz}{d(\ \ )}$ (125)

Equation 125 appears formidable until it is taken from the general form shown to a specific form. For example, putting x in the bracket of Equation 125 results in:

$$\frac{dy}{dx} = \frac{\partial y}{\partial x} \frac{dx}{dx} + \frac{\partial y}{\partial z} \frac{dz}{dx}$$

or

$$\frac{dy}{dx} = \frac{\partial y}{\partial x} + \frac{\partial y}{\partial z} \frac{dz}{dx} \qquad (126)$$

Equation 126 states that dy/dx is the same as $\partial y/\partial x$ (which assumed that z remained constant) plus a second term in the event that there is a change in z when x changes. If z does not change as x changes, then $dz/dx = 0$ and Equation

126 reduces to:

$$\frac{dy}{dx} = \frac{\partial y}{\partial x}$$

A common problem is one in which both x and z are functions in turn of another variable like time, t. The derivative formula (Equation 125) now becomes

$$\frac{dy}{dt} = \frac{\partial y}{\partial x}\frac{dx}{dt} + \frac{\partial y}{\partial z}\frac{dz}{dt}$$

This formula says that the rate at which y changes with respect to changes in time equals the sum of two terms: (1) the rate at which x changes with changes in time multiplied by the rate at which y changes with respect to x, (i.e. the rate at which y changes due to time changes affecting y through x) and (2) the rate at which y changes due to time changes affecting y through z. These formulas can be extended to cases of many interrelated variables but their basic meanings remain the same.

13. DIFFERENTIAL EQUATIONS II

NOMENCLATURE. Differential equations can be divided into two classes—*ordinary* differential equations which have only one independent variable and *partial* differential equations which have more than one independent variable. Only ordinary differential equations are discussed in this book. All the equations discussed up to this chapter have been *linear* differential equations, i.e., equations in which derivatives have neither been raised to a power nor multiplied by each other. In addition, all equations have had derivative terms with constant coefficients.

CHANGE OF VARIABLES. Four methods for solving ordinary linear differential equations with constant coefficients were discussed in Chapter 6. All four methods were illustrated with time as the independent variable and by a simple change of variables time can be made to be the independent variable in any ordinary differential equation. For example, such a problem as:

$$\frac{d^2y}{dz^2} + 7\frac{dy}{dz} - 14y = 2\sin z$$

becomes solvable by the methods of Chapter 6 when a change of variables via $z = t$ is used to generate:

$$\frac{d^2y}{dt^2} + 7\frac{dy}{dt} - 14y = 2\sin t$$

There is another general method often used to solve differential equations by changes of variables which might be called the historic method. It consists of finding a differential equation that someone has solved in the past and performing one or more change of variables operations on the previously developed differential equation and its solution until the form of the present problem differential equation and its solution appear. This technique has been developed by some mathematicians as a fine art and the historic equations that find such use are often named after the man who first solved them. There are so many analogous problems in various branches of physics and engineering that this method has been quite useful in the past. The increasing use of digital and analog computers, as discussed later in this chapter, is reducing the demand for this art.

SIMULTANEOUS EQUATIONS. Sometimes the statement of a problem includes several dependent variables in the form of simultaneous differential equations. As an example,

$$\frac{d^2y}{dt^2} + 3\frac{dz}{dt} + 4y = 0 \tag{127}$$

$$\frac{d^2z}{dt^2} + 7\frac{dy}{dt} + z = 2t \tag{128}$$

(Boundary conditions: z, y and all their derivatives are zero at $t = 0$.)

The combination of Equations 127 and 128 is called a set of simultaneous equations and for the problem to be solvable there must be as many equations in the set as there are dependent variables (in this case, too).

In general the classical method cannot be readily used to solve simultaneous differential equations but the other three methods discussed in Chapter 6 can be used. (It is suggested that the reader review Chapter 6 at this time.)

Figure 69 shows the electronic analog computer set-up for solving Equations 127 and 128. Notice that for each equation there is a summing device which generates the highest order derivative of one of the dependent variables that appears in the problem and that by integration the lower order derivatives are generated so that they can provide inputs to the summing networks that generate the higher order derivatives. If there are any initial conditions in the problem they are inserted by causing the outputs of the integrators to have such values at time $t = 0^+$.

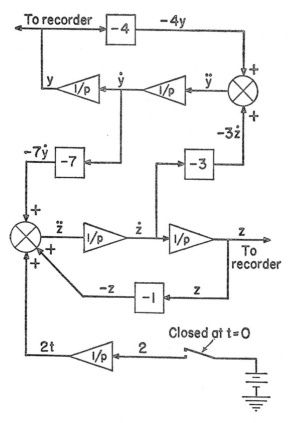

Fig. 69

Laplace operational calculus can be used to solve simultaneous differential equations. The transformation is made in the usual way to yield a set of simultaneous algebraic equations. In the case of Equations 127 and 128 the set of transformed equations is:

$$p^2 Y + 3pZ + 4Y = 0$$
$$p^2 Z + 7pY + Z = 2/p^2$$

The transformed equations are solved by algebra to develop separate equations for each of the transformed dependent variables (Y and Z in this case) as functions of p. These separate equations are then inverse transformed to yield the desired solutions, y and z as functions of t.

By using digital computers, step-by-step numerical methods can be used to solve simultaneous differential equations by a scheme that is similar to the electronic analog computer technique. These computers are discussed in the next section.

SOLVING NON-LINEAR DIFFERENTIAL EQUATIONS. Non-linear differential equations can be solved by either analog or digital computer techniques. An example will be used to illustrate the digital method using a digital differential analyzer. If the problem is to solve:

$$\frac{d^3z}{dy^3} - \left(\frac{d^2z}{dy^2}\right)\left(\frac{dz}{dy}\right) + 2y^3 = 7$$

a computer is programmed as shown in Fig. 70.

As in the case of the electronic analog computer, the program is fundamentally generated by assuming that the highest derivative (d^3z/dy^3) is known, integrating to generate lower order derivatives, and then using the lower order derivatives so generated to form the highest derivative that was originally assumed to be known. Integration in each integration block consists of adding the input signal of the block to the sum in a storage register each time that the clock generates a Δy pulse. Thus the storage register in an integrator keeps a running tally of the area under the curve of the input plotted against y; hence the tally of the register is the integral of the input and is used as the output of the integrator block. The block labeled "π" is a multiplier whose output is the product of its inputs. The block labeled "function generator" contains a stored program of the desired function (y^3 vs y) and generates the required output for any input. Both y and z signals go to a recorder to display the problem solution. Since the computation is done step by step as the clock generates Δy pulses, this is the same basic technique used in the step-by-step differential equation solution method of Chapter 6, but with enough flexibility and automation to permit the ready computation of non-linear differential equations.

Electronic analog computers are available with multiplication devices and non-linear function generators so that

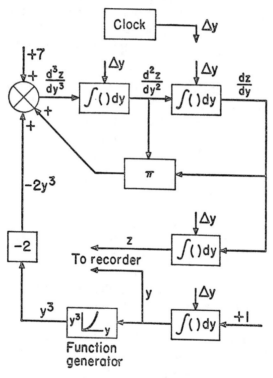

Fig. 70

they too can be used to solve non-linear differential equations. In general the electronic analog computer is easier to program but less accurate than a digital computer.

Some non-linear differential equations can be easily solved by Laplace operational calculus, for example those whose non-linearities are a result of multiplications by factors of t in the equation to be solved.

APPENDIX

TABLE I

SOME BASIC DERIVATIVES

$$\frac{dc}{dx} = 0 \quad (c \text{ is any constant})$$

$$\frac{d(x_n^n)}{dx} = nx^{n-1}$$

$$\frac{d(\sin x)}{dx} = \cos x$$

$$\frac{d(\cos x)}{dx} = -\sin x$$

$$\frac{de^x}{dx} = e^x$$

TABLE II

DERIVATIVES OF INVOLVED EXPRESSIONS

Rule 1. $\dfrac{d(au)}{dx} = a\,\dfrac{du}{dx}$

Rule 2. $\dfrac{d(u + v)}{dx} = \dfrac{du}{dx} + \dfrac{dv}{dx}$

Rule 3. $\dfrac{d(u \cdot v)}{dx} = u\,\dfrac{dv}{dx} + v\,\dfrac{du}{dx}$

Rule 4. $\dfrac{d(u \div v)}{dx} = \dfrac{1}{v^2}\left(v\,\dfrac{du}{dx} - u\,\dfrac{dv}{dx}\right)$

$$a = \text{constant}$$
$$u = f_1(x)$$
$$v = f_2(x)$$

TABLE III

COMMONLY USED DERIVATIVES

$$\int x^n \, dx = \frac{x^{n+1}}{n+1} + c \quad (n \neq -1)$$

$$\int \frac{dx}{x} = \log_e x + c$$

$$\int a^x \, dx = \frac{a^x}{\log_e a} + c$$

$$\int e^x \, dx = e^x + c$$

$$\int \sin x \, dx = -\cos x + c$$

$$\int \cos x \, dx = \sin x + c$$

$$\int \sin^2 x \, dx = \tfrac{1}{2}(x - \sin x \cos x) + c$$

$$\int \cos^2 x \, dx = \tfrac{1}{2}(x + \sin x \cos x) + c$$

$$\int e^{ax} \, dx = \frac{e^{ax}}{a} + c$$

x is a variable; n is an integer; c and a are numbers.

TABLE IV

LAWS OF INTEGRALS

Rule 1. $\displaystyle\int az\,dx = a\int z\,dx$

Rule 2. $\displaystyle\int (z + y)\,dx = \int z\,dx + \int y\,dx$

Rule 3. $\displaystyle\int_{x=a}^{x=c} z\,dx = \int_{x=a}^{x=b} z\,dx + \int_{x=b}^{x=c} z\,dx$

Rule 4. $\displaystyle\int_{x=a}^{x=b} z\,dx = -\int_{x=b}^{x=a} z\,dx$

Symbols x, y, and z are variables. a is a number.

TABLE V

COMMONLY USED SYMBOLS FOR DIFFERENTIATION
AND INTEGRATION WITH RESPECT TO TIME

Traditional form	x	$\dfrac{dx}{dt}$	$\dfrac{d^2x}{dt^2}$	$\dfrac{d^3x}{dt^3}$	$\displaystyle\int x\,dt$
Dot notation	x	$\dot{x}$	$\ddot{x}$	$\dddot{x}$	$\displaystyle\int x\,dt$
Heaviside operational notation	x	Px	P^2x	P^3x	$(1/P)x$ or $P^{-1}x$
Alternate operational notation	x	Sx	S^2x	S^3x	x/S or $S^{-1}x$
Advanced calculus notation	x	$D(x)$	$D^2(x)$	$D^3(x)$	$D^{-1}(x)$
Function notation	$f(t)$	$f'(t)$	$f''(t)$	$f'''(t)$	$f^{-1}(t)$

TABLE VI

ELEMENTARY LAPLACE TRANSFORMS

Real Quantity	Transformed Quantity
V_2	V_{2p}
$\dot{V}_2$	$pV_{2p} - V_2(0)$
1	$1/p$
t	$1/p^2$
e^{-at}	$\dfrac{1}{p+a}$

TABLE VII

LAPLACE TRANSFORMS

Time Function (x)	Laplace Transform (X)
1	$\dfrac{1}{p}$
unit step at $t = 0^+$	$\dfrac{1}{p}$
e^{-kt}	$\dfrac{1}{p+k}$
t	$\dfrac{1}{p^2}$
t^2	$\dfrac{2}{p^3}$
t^n	$\dfrac{n!}{p^{(n+1)}}$
te^{-kt}	$\dfrac{1}{(p+k)^2}$
$\sin kt$	$\dfrac{k}{p^2 + k^2}$
$\cos kt$	$\dfrac{p}{p^2 + k^2}$
$\dfrac{1}{(n-1)!}\, t^{(n-1)}e^{-kt}$	$\dfrac{1}{(p+k)^n}$
$e^{-at}(E\cos kt + F\sin kt)$	$\dfrac{E(p+a) + Fk}{(p+a)^2 + k^2}$

TABLE VIII

LAPLACE TRANSFORM THEOREMS

Time Function	Laplace Transform	Notes
x	X	notation
$\dfrac{dx}{dt}$	$pX - (x)_{0^+}$*	first derivative
$\dfrac{d^2x}{dt^2}$	$p^2X - p(x)_{0^+} - (dx/dt)_{0^+}$	second derivative
$\dfrac{d^nx}{dt^n}$	$\begin{bmatrix} p^nX - p^{(n-1)}(x)_{0^+} \\ - p^{(n-2)}(dx/dt)_{0^+} \\ - p^{(n-3)}(d^2x/dt^2)_{0^+} \\ - \cdots - \\ (d^{(n-1)}(x)/dt^{(n-1)})_{0^+} \end{bmatrix}$	general derivative
$\displaystyle\int_{-\infty}^{t} x\, dt$	$\dfrac{X + \left[\displaystyle\int_{-\infty}^{0} x\, dt\right]}{p}$	integral
$\displaystyle\int_{-\infty}^{t} x\, dt$	$\dfrac{X}{p}$	integral with zero constant of integration
$\displaystyle\int_{0}^{t}\left[\int_{0}^{t} x\, dt\right] dt$	$\dfrac{X}{p^2}$	repeat of above
$e^{-kt}x$	$X(p + k)$	shifting theorem
$x(t - t_1)$	$Xe^{-t_1 p}$	time delay†
$x + y$	$X + Y$	addition of transforms
kx	kX	multiplication by a constant
$\displaystyle\lim_{t \to \infty} (x)$	$= \displaystyle\lim_{p \to 0} (pX)$	final value†
$\displaystyle\lim_{t \to 0} (x)$	$= \displaystyle\lim_{p \to \infty} (pX)$	initial value

* 0^+ means "at $t = 0^+$."

† See text for restrictions.

INDEX

A CATALOG OF SELECTED

DOVER BOOKS

IN SCIENCE AND MATHEMATICS

Astronomy

BURNHAM'S CELESTIAL HANDBOOK, Robert Burnham, Jr. Thorough guide to the stars beyond our solar system. Exhaustive treatment. Alphabetical by constellation: Andromeda to Cetus in Vol. 1; Chamaeleon to Orion in Vol. 2; and Pavo to Vulpecula in Vol. 3. Hundreds of illustrations. Index in Vol. 3. 2,000pp. 6⅛ x 9¼.
23567-X, 23568-8, 23673-0 Three-vol. set

THE EXTRATERRESTRIAL LIFE DEBATE, 1750–1900, Michael J. Crowe. First detailed, scholarly study in English of the many ideas that developed from 1750 to 1900 regarding the existence of intelligent extraterrestrial life. Examines ideas of Kant, Herschel, Voltaire, Percival Lowell, many other scientists and thinkers. 16 illustrations. 704pp. 5⅜ x 8½. 40675-X

A HISTORY OF ASTRONOMY, A. Pannekoek. Well-balanced, carefully reasoned study covers such topics as Ptolemaic theory, work of Copernicus, Kepler, Newton, Eddington's work on stars, much more. Illustrated. References. 521pp. 5⅜ x 8½.
65994-1

AMATEUR ASTRONOMER'S HANDBOOK, J. B. Sidgwick. Timeless, comprehensive coverage of telescopes, mirrors, lenses, mountings, telescope drives, micrometers, spectroscopes, more. 189 illustrations. 576pp. 5⅝ x 8¼. (Available in U.S. only.)
24034-7

STARS AND RELATIVITY, Ya. B. Zel'dovich and I. D. Novikov. Vol. 1 of *Relativistic Astrophysics* by famed Russian scientists. General relativity, properties of matter under astrophysical conditions, stars, and stellar systems. Deep physical insights, clear presentation. 1971 edition. References. 544pp. 5⅝ x 8¼. 69424-0

Chemistry

CHEMICAL MAGIC, Leonard A. Ford. Second Edition, Revised by E. Winston Grundmeier. Over 100 unusual stunts demonstrating cold fire, dust explosions, much more. Text explains scientific principles and stresses safety precautions. 128pp. 5⅜ x 8½. 67628-5

THE DEVELOPMENT OF MODERN CHEMISTRY, Aaron J. Ihde. Authoritative history of chemistry from ancient Greek theory to 20th-century innovation. Covers major chemists and their discoveries. 209 illustrations. 14 tables. Bibliographies. Indices. Appendices. 851pp. 5⅜ x 8½. 64235-6

CATALYSIS IN CHEMISTRY AND ENZYMOLOGY, William P. Jencks. Exceptionally clear coverage of mechanisms for catalysis, forces in aqueous solution, carbonyl- and acyl-group reactions, practical kinetics, more. 864pp. 5⅜ x 8½.
65460-5

THE HISTORICAL BACKGROUND OF CHEMISTRY, Henry M. Leicester. Evolution of ideas, not individual biography. Concentrates on formulation of a coherent set of chemical laws. 260pp. 5⅜ x 8½. 61053-5

A SHORT HISTORY OF CHEMISTRY, J. R. Partington. Classic exposition explores origins of chemistry, alchemy, early medical chemistry, nature of atmosphere, theory of valency, laws and structure of atomic theory, much more. 428pp. 5⅜ x 8½. (Available in U.S. only.) 65977-1

GENERAL CHEMISTRY, Linus Pauling. Revised 3rd edition of classic first-year text by Nobel laureate. Atomic and molecular structure, quantum mechanics, statistical mechanics, thermodynamics correlated with descriptive chemistry. Problems. 992pp. 5⅜ x 8½. 65622-5

Engineering

DE RE METALLICA, Georgius Agricola. The famous Hoover translation of greatest treatise on technological chemistry, engineering, geology, mining of early modern times (1556). All 289 original woodcuts. 638pp. 6¾ x 11. 60006-8

FUNDAMENTALS OF ASTRODYNAMICS, Roger Bate et al. Modern approach developed by U.S. Air Force Academy. Designed as a first course. Problems, exercises. Numerous illustrations. 455pp. 5⅜ x 8½. 60061-0

DYNAMICS OF FLUIDS IN POROUS MEDIA, Jacob Bear. For advanced students of ground water hydrology, soil mechanics and physics, drainage and irrigation engineering and more. 335 illustrations. Exercises, with answers. 784pp. 6⅛ x 9¼. 65675-6

ANALYTICAL MECHANICS OF GEARS, Earle Buckingham. Indispensable reference for modern gear manufacture covers conjugate gear-tooth action, gear-tooth profiles of various gears, many other topics. 263 figures. 102 tables. 546pp. 5⅜ x 8½. 65712-4

MECHANICS, J. P. Den Hartog. A classic introductory text or refresher. Hundreds of applications and design problems illuminate fundamentals of trusses, loaded beams and cables, etc. 334 answered problems. 462pp. 5⅜ x 8½. 60754-2

MECHANICAL VIBRATIONS, J. P. Den Hartog. Classic textbook offers lucid explanations and illustrative models, applying theories of vibrations to a variety of practical industrial engineering problems. Numerous figures. 233 problems, solutions. Appendix. Index. Preface. 436pp. 5⅜ x 8½. 64785-4

STRENGTH OF MATERIALS, J. P. Den Hartog. Full, clear treatment of basic material (tension, torsion, bending, etc.) plus advanced material on engineering methods, applications. 350 answered problems. 323pp. 5⅜ x 8½. 60755-0

A HISTORY OF MECHANICS, René Dugas. Monumental study of mechanical principles from antiquity to quantum mechanics. Contributions of ancient Greeks, Galileo, Leonardo, Kepler, Lagrange, many others. 671pp. 5⅜ x 8½. 65632-2

METAL FATIGUE, N. E. Frost, K. J. Marsh, and L. P. Pook. Definitive, clearly written, and well-illustrated volume addresses all aspects of the subject, from the historical development of understanding metal fatigue to vital concepts of the cyclic stress that causes a crack to grow. Includes 7 appendixes. 544pp. 5⅜ x 8½. 40927-9

STATISTICAL MECHANICS: Principles and Applications, Terrell L. Hill. Standard text covers fundamentals of statistical mechanics, applications to fluctuation theory, imperfect gases, distribution functions, more. 448pp. 5⅜ x 8½. 65390-0

THE VARIATIONAL PRINCIPLES OF MECHANICS, Cornelius Lanczos. Graduate level coverage of calculus of variations, equations of motion, relativistic mechanics, more. First inexpensive paperbound edition of classic treatise. Index. Bibliography. 418pp. 5⅜ x 8½. 65067-7

THE VARIOUS AND INGENIOUS MACHINES OF AGOSTINO RAMELLI: A Classic Sixteenth-Century Illustrated Treatise on Technology, Agostino Ramelli. One of the most widely known and copied works on machinery in the 16th century. 194 detailed plates of water pumps, grain mills, cranes, more. 608pp. 9 x 12. 28180-9

ORDINARY DIFFERENTIAL EQUATIONS AND STABILITY THEORY: An Introduction, David A. Sánchez. Brief, modern treatment. Linear equation, stability theory for autonomous and nonautonomous systems, etc. 164pp. 5⅜ x 8¼. 63828-6

ROTARY WING AERODYNAMICS, W. Z. Stepniewski. Clear, concise text covers aerodynamic phenomena of the rotor and offers guidelines for helicopter performance evaluation. Originally prepared for NASA. 537 figures. 640pp. 6⅛ x 9¼. 64647-5

INTRODUCTION TO SPACE DYNAMICS, William Tyrrell Thomson. Comprehensive, classic introduction to space-flight engineering for advanced undergraduate and graduate students. Includes vector algebra, kinematics, transformation of coordinates. Bibliography. Index. 352pp. 5⅜ x 8½. 65113-4

HISTORY OF STRENGTH OF MATERIALS, Stephen P. Timoshenko. Excellent historical survey of the strength of materials with many references to the theories of elasticity and structure. 245 figures. 452pp. 5⅜ x 8½. 61187-6

ANALYTICAL FRACTURE MECHANICS, David J. Unger. Self-contained text supplements standard fracture mechanics texts by focusing on analytical methods for determining crack-tip stress and strain fields. 336pp. 6⅛ x 9¼. 41737-9

Mathematics

HANDBOOK OF MATHEMATICAL FUNCTIONS WITH FORMULAS, GRAPHS, AND MATHEMATICAL TABLES, edited by Milton Abramowitz and Irene A. Stegun. Vast compendium: 29 sets of tables, some to as high as 20 places. 1,046pp. 8 x 10½. 61272-4

FUNCTIONAL ANALYSIS (Second Corrected Edition), George Bachman and Lawrence Narici. Excellent treatment of subject geared toward students with background in linear algebra, advanced calculus, physics and engineering. Text covers introduction to inner-product spaces, normed, metric spaces, and topological spaces; complete orthonormal sets, the Hahn-Banach Theorem and its consequences, and many other related subjects. 1966 ed. 544pp. 6⅛ x 9¼. 40251-7

ASYMPTOTIC EXPANSIONS OF INTEGRALS, Norman Bleistein & Richard A. Handelsman. Best introduction to important field with applications in a variety of scientific disciplines. New preface. Problems. Diagrams. Tables. Bibliography. Index. 448pp. 5⅜ x 8½. 65082-0

FAMOUS PROBLEMS OF GEOMETRY AND HOW TO SOLVE THEM, Benjamin Bold. Squaring the circle, trisecting the angle, duplicating the cube: learn their history, why they are impossible to solve, then solve them yourself. 128pp. 5⅜ x 8½. 24297-8

VECTOR AND TENSOR ANALYSIS WITH APPLICATIONS, A. I. Borisenko and I. E. Tarapov. Concise introduction. Worked-out problems, solutions, exercises. 257pp. 5⅝ x 8¼. 63833-2

THE ABSOLUTE DIFFERENTIAL CALCULUS (CALCULUS OF TENSORS), Tullio Levi-Civita. Great 20th-century mathematician's classic work on material necessary for mathematical grasp of theory of relativity. 452pp. 5⅝ x 8¼. 63401-9

AN INTRODUCTION TO ORDINARY DIFFERENTIAL EQUATIONS, Earl A. Coddington. A thorough and systematic first course in elementary differential equations for undergraduates in mathematics and science, with many exercises and problems (with answers). Index. 304pp. 5⅜ x 8½. 65942-9

FOURIER SERIES AND ORTHOGONAL FUNCTIONS, Harry F. Davis. An incisive text combining theory and practical example to introduce Fourier series, orthogonal functions and applications of the Fourier method to boundary-value problems. 570 exercises. Answers and notes. 416pp. 5⅜ x 8½. 65973-9

COMPUTABILITY AND UNSOLVABILITY, Martin Davis. Classic graduate-level introduction to theory of computability, usually referred to as theory of recurrent functions. New preface and appendix. 288pp. 5⅜ x 8½. 61471-9

ASYMPTOTIC METHODS IN ANALYSIS, N. G. de Bruijn. An inexpensive, comprehensive guide to asymptotic methods–the pioneering work that teaches by explaining worked examples in detail. Index. 224pp. 5⅜ x 8½ 64221-6

ESSAYS ON THE THEORY OF NUMBERS, Richard Dedekind. Two classic essays by great German mathematician: on the theory of irrational numbers; and on transfinite numbers and properties of natural numbers. 115pp. 5⅜ x 8½. 21010-3

APPLIED COMPLEX VARIABLES, John W. Dettman. Step-by-step coverage of fundamentals of analytic function theory–plus lucid exposition of five important applications: Potential Theory; Ordinary Differential Equations; Fourier Transforms; Laplace Transforms; Asymptotic Expansions. 66 figures. Exercises at chapter ends. 512pp. 5⅜ x 8½. 64670-X

INTRODUCTION TO LINEAR ALGEBRA AND DIFFERENTIAL EQUATIONS, John W. Dettman. Excellent text covers complex numbers, determinants, orthonormal bases, Laplace transforms, much more. Exercises with solutions. Undergraduate level. 416pp. 5⅜ x 8½. 65191-6

MATHEMATICAL METHODS IN PHYSICS AND ENGINEERING, John W. Dettman. Algebraically based approach to vectors, mapping, diffraction, other topics in applied math. Also generalized functions, analytic function theory, more. Exercises. 448pp. 5⅝ x 8¼. 65649-7

CALCULUS OF VARIATIONS WITH APPLICATIONS, George M. Ewing. Applications-oriented introduction to variational theory develops insight and promotes understanding of specialized books, research papers. Suitable for advanced undergraduate/graduate students as primary, supplementary text. 352pp. 5⅜ x 8½. 64856-7

COMPLEX VARIABLES, Francis J. Flanigan. Unusual approach, delaying complex algebra till harmonic functions have been analyzed from real variable viewpoint. Includes problems with answers. 364pp. 5⅜ x 8½. 61388-7

AN INTRODUCTION TO THE CALCULUS OF VARIATIONS, Charles Fox. Graduate-level text covers variations of an integral, isoperimetrical problems, least action, special relativity, approximations, more. References. 279pp. 5⅜ x 8½. 65499-0

CATASTROPHE THEORY FOR SCIENTISTS AND ENGINEERS, Robert Gilmore. Advanced-level treatment describes mathematics of theory grounded in the work of Poincaré, R. Thom, other mathematicians. Also important applications to problems in mathematics, physics, chemistry and engineering. 1981 edition. References. 28 tables. 397 black-and-white illustrations. xvii + 666pp. 6⅛ x 9¼. 67539-4

INTRODUCTION TO DIFFERENCE EQUATIONS, Samuel Goldberg. Exceptionally clear exposition of important discipline with applications to sociology, psychology, economics. Many illustrative examples; over 250 problems. 260pp. 5⅜ x 8½. 65084-7

NUMERICAL METHODS FOR SCIENTISTS AND ENGINEERS, Richard Hamming. Classic text stresses frequency approach in coverage of algorithms, polynomial approximation, Fourier approximation, exponential approximation, other topics. Revised and enlarged 2nd edition. 721pp. 5⅜ x 8½. 65241-6

INTRODUCTION TO NUMERICAL ANALYSIS (2nd Edition), F. B. Hildebrand. Classic, fundamental treatment covers computation, approximation, interpolation, numerical differentiation and integration, other topics. 150 new problems. 669pp. 5⅜ x 8½. 65363-3

THE FUNCTIONS OF MATHEMATICAL PHYSICS, Harry Hochstadt. Comprehensive treatment of orthogonal polynomials, hypergeometric functions, Hill's equation, much more. Bibliography. Index. 322pp. 5⅜ x 8½.   65214-9

THREE PEARLS OF NUMBER THEORY, A. Y. Khinchin. Three compelling puzzles require proof of a basic law governing the world of numbers. Challenges concern van der Waerden's theorem, the Landau-Schnirelmann hypothesis and Mann's theorem, and a solution to Waring's problem. Solutions included. 64pp. 5⅜ x 8½.
40026-3

CALCULUS REFRESHER FOR TECHNICAL PEOPLE, A. Albert Klaf. Covers important aspects of integral and differential calculus via 756 questions. 566 problems, most answered. 431pp. 5⅜ x 8½.   20370-0

THE PHILOSOPHY OF MATHEMATICS: An Introductory Essay, Stephan Körner. Surveys the views of Plato, Aristotle, Leibniz & Kant concerning propositions and theories of applied and pure mathematics. Introduction. Two appendices. Index. 198pp. 5⅜ x 8½.   25048-2

INTRODUCTORY REAL ANALYSIS, A.N. Kolmogorov, S. V. Fomin. Translated by Richard A. Silverman. Self-contained, evenly paced introduction to real and functional analysis. Some 350 problems. 403pp. 5⅜ x 8½.   61226-0

APPLIED ANALYSIS, Cornelius Lanczos. Classic work on analysis and design of finite processes for approximating solution of analytical problems. Algebraic equations, matrices, harmonic analysis, quadrature methods, much more. 559pp. 5⅜ x 8½.
65656-X

AN INTRODUCTION TO ALGEBRAIC STRUCTURES, Joseph Landin. Superb self-contained text covers "abstract algebra": sets and numbers, theory of groups, theory of rings, much more. Numerous well-chosen examples, exercises. 247pp. 5⅜ x 8½.
65940-2

SPECIAL FUNCTIONS, N. N. Lebedev. Translated by Richard Silverman. Famous Russian work treating more important special functions, with applications to specific problems of physics and engineering. 38 figures. 308pp. 5⅜ x 8½.   60624-4

QUALITATIVE THEORY OF DIFFERENTIAL EQUATIONS, V. V. Nemytskii and V.V. Stepanov. Classic graduate-level text by two prominent Soviet mathematicians covers classical differential equations as well as topological dynamics and ergodic theory. Bibliographies. 523pp. 5⅜ x 8½.   65954-2

NUMBER THEORY AND ITS HISTORY, Oystein Ore. Unusually clear, accessible introduction covers counting, properties of numbers, prime numbers, much more. Bibliography. 380pp. 5⅜ x 8½.   65620-9

THEORY OF MATRICES, Sam Perlis. Outstanding text covering rank, nonsingularity and inverses in connection with the development of canonical matrices under the relation of equivalence, and without the intervention of determinants. Includes exercises. 237pp. 5⅜ x 8½.   66810-X

INTRODUCTION TO ANALYSIS, Maxwell Rosenlicht. Unusually clear, accessible coverage of set theory, real number system, metric spaces, continuous functions, Riemann integration, multiple integrals, more. Wide range of problems. Undergraduate level. Bibliography. 254pp. 5⅜ x 8½. 65038-3

MODERN NONLINEAR EQUATIONS, Thomas L. Saaty. Emphasizes practical solution of problems; covers seven types of equations. ". . . a welcome contribution to the existing literature...."–*Math Reviews*. 490pp. 5⅜ x 8½. 64232-1

MATRICES AND LINEAR ALGEBRA, Hans Schneider and George Phillip Barker. Basic textbook covers theory of matrices and its applications to systems of linear equations and related topics such as determinants, eigenvalues and differential equations. Numerous exercises. 432pp. 5⅜ x 8½. 66014-1

MATHEMATICS APPLIED TO CONTINUUM MECHANICS, Lee A. Segel. Analyzes models of fluid flow and solid deformation. For upper-level math, science and engineering students. 608pp. 5⅜ x 8½. 65369-2

ELEMENTS OF REAL ANALYSIS, David A. Sprecher. Classic text covers fundamental concepts, real number system, point sets, functions of a real variable, Fourier series, much more. Over 500 exercises. 352pp. 5⅜ x 8½. 65385-4

AN INTRODUCTION TO MATRICES, SETS AND GROUPS FOR SCIENCE STUDENTS, G. Stephenson. Concise, readable text introduces sets, groups, and most importantly, matrices to undergraduate students of physics, chemistry, and engineering. Problems. 164pp. 5⅜ x 8½. 65077-4

SET THEORY AND LOGIC, Robert R. Stoll. Lucid introduction to unified theory of mathematical concepts. Set theory and logic seen as tools for conceptual understanding of real number system. 496pp. 5⅜ x 8¼. 63829-4

TENSOR CALCULUS, J.L. Synge and A. Schild. Widely used introductory text covers spaces and tensors, basic operations in Riemannian space, non-Riemannian spaces, etc. 324pp. 5⅜ x 8¼. 63612-7

ORDINARY DIFFERENTIAL EQUATIONS, Morris Tenenbaum and Harry Pollard. Exhaustive survey of ordinary differential equations for undergraduates in mathematics, engineering, science. Thorough analysis of theorems. Diagrams. Bibliography. Index. 818pp. 5⅜ x 8½. 64940-7

INTEGRAL EQUATIONS, F. G. Tricomi. Authoritative, well-written treatment of extremely useful mathematical tool with wide applications. Volterra Equations, Fredholm Equations, much more. Advanced undergraduate to graduate level. Exercises. Bibliography. 238pp. 5⅜ x 8½. 64828-1

FOURIER SERIES, Georgi P. Tolstov. Translated by Richard A. Silverman. A valuable addition to the literature on the subject, moving clearly from subject to subject and theorem to theorem. 107 problems, answers. 336pp. 5⅜ x 8½. 63317-9

POPULAR LECTURES ON MATHEMATICAL LOGIC, Hao Wang. Noted logician's lucid treatment of historical developments, set theory, model theory, recursion theory and constructivism, proof theory, more. 3 appendixes. Bibliography. 1981 edition. ix + 283pp. 5⅜ x 8½.   67632-3

CALCULUS OF VARIATIONS, Robert Weinstock. Basic introduction covering isoperimetric problems, theory of elasticity, quantum mechanics, electrostatics, etc. Exercises throughout. 326pp. 5⅜ x 8½.   63069-2

THE CONTINUUM: A Critical Examination of the Foundation of Analysis, Hermann Weyl. Classic of 20th-century foundational research deals with the conceptual problem posed by the continuum. 156pp. 5⅜ x 8½.   67982-9

CHALLENGING MATHEMATICAL PROBLEMS WITH ELEMENTARY SOLUTIONS, A. M. Yaglom and I. M. Yaglom. Over 170 challenging problems on probability theory, combinatorial analysis, points and lines, topology, convex polygons, many other topics. Solutions. Total of 445pp. 5⅜ x 8½. Two-vol. set.
Vol. I: 65536-9 Vol. II: 65537-7

A SURVEY OF NUMERICAL MATHEMATICS, David M. Young and Robert Todd Gregory. Broad self-contained coverage of computer-oriented numerical algorithms for solving various types of mathematical problems in linear algebra, ordinary and partial, differential equations, much more. Exercises. Total of 1,248pp. 5⅜ x 8½.
Two volumes.   Vol. I: 65691-8 Vol. II: 65692-6

INTRODUCTION TO PARTIAL DIFFERENTIAL EQUATIONS WITH APPLICATIONS, E. C. Zachmanoglou and Dale W. Thoe. Essentials of partial differential equations applied to common problems in engineering and the physical sciences. Problems and answers. 416pp. 5⅜ x 8½.   65251-3

THE THEORY OF GROUPS, Hans J. Zassenhaus. Well-written graduate-level text acquaints reader with group-theoretic methods and demonstrates their usefulness in mathematics. Axioms, the calculus of complexes, homomorphic mapping, p-group theory, more. Many proofs shorter and more transparent than older ones. 276pp. 5⅜ x 8½.   40922-8

DISTRIBUTION THEORY AND TRANSFORM ANALYSIS: An Introduction to Generalized Functions, with Applications, A. H. Zemanian. Provides basics of distribution theory, describes generalized Fourier and Laplace transformations. Numerous problems. 384pp. 5⅜ x 8½.   65479-6

Math–Decision Theory, Statistics, Probability

ELEMENTARY DECISION THEORY, Herman Chernoff and Lincoln E. Moses. Clear introduction to statistics and statistical theory covers data processing, probability and random variables, testing hypotheses, much more. Exercises. 364pp. 5⅜ x 8½.   65218-1

CATALOG OF DOVER BOOKS

STATISTICS MANUAL, Edwin L. Crow et al. Comprehensive, practical collection of classical and modern methods prepared by U.S. Naval Ordnance Test Station. Stress on use. Basics of statistics assumed. 288pp. 5⅜ x 8½. 60599-X

SOME THEORY OF SAMPLING, William Edwards Deming. Analysis of the problems, theory and design of sampling techniques for social scientists, industrial managers and others who find statistics important at work. 61 tables. 90 figures. xvii +602pp. 5⅜ x 8½. 64684-X

STATISTICAL ADJUSTMENT OF DATA, W. Edwards Deming. Introduction to basic concepts of statistics, curve fitting, least squares solution, conditions without parameter, conditions containing parameters. 26 exercises worked out. 271pp. 5⅜ x 8½. 64685-8

LINEAR PROGRAMMING AND ECONOMIC ANALYSIS, Robert Dorfman, Paul A. Samuelson and Robert M. Solow. First comprehensive treatment of linear programming in standard economic analysis. Game theory, modern welfare economics, Leontief input-output, more. 525pp. 5⅜ x 8½. 65491-5

DICTIONARY/OUTLINE OF BASIC STATISTICS, John E. Freund and Frank J. Williams. A clear concise dictionary of over 1,000 statistical terms and an outline of statistical formulas covering probability, nonparametric tests, much more. 208pp. 5⅜ x 8½. 66796-0

PROBABILITY: An Introduction, Samuel Goldberg. Excellent basic text covers set theory, probability theory for finite sample spaces, binomial theorem, much more. 360 problems. Bibliographies. 322pp. 5⅜ x 8½. 65252-1

GAMES AND DECISIONS: Introduction and Critical Survey, R. Duncan Luce and Howard Raiffa. Superb nontechnical introduction to game theory, primarily applied to social sciences. Utility theory, zero-sum games, n-person games, decision-making, much more. Bibliography. 509pp. 5⅜ x 8½. 65943-7

FIFTY CHALLENGING PROBLEMS IN PROBABILITY WITH SOLUTIONS, Frederick Mosteller. Remarkable puzzlers, graded in difficulty, illustrate elementary and advanced aspects of probability. Detailed solutions. 88pp. 5⅜ x 8½. 65355-2

PROBABILITY THEORY: A Concise Course, Y. A. Rozanov. Highly readable, self-contained introduction covers combination of events, dependent events, Bernoulli trials, etc. 148pp. 5⅜ x 8¼. 63544-9

STATISTICAL METHOD FROM THE VIEWPOINT OF QUALITY CONTROL, Walter A. Shewhart. Important text explains regulation of variables, uses of statistical control to achieve quality control in industry, agriculture, other areas. 192pp. 5⅜ x 8½. 65232-7

THE COMPLEAT STRATEGYST: Being a Primer on the Theory of Games of Strategy, J. D. Williams. Highly entertaining classic describes, with many illustrated examples, how to select best strategies in conflict situations. Prefaces. Appendices. 268pp. 5⅜ x 8½. 25101-2

Math–Geometry and Topology

ELEMENTARY CONCEPTS OF TOPOLOGY, Paul Alexandroff. Elegant, intuitive approach to topology from set-theoretic topology to Betti groups; how concepts of topology are useful in math and physics. 25 figures. 57pp. 5⅜ x 8½. 60747-X

COMBINATORIAL TOPOLOGY, P. S. Alexandrov. Clearly written, well-organized, three-part text begins by dealing with certain classic problems without using the formal techniques of homology theory and advances to the central concept, the Betti groups. Numerous detailed examples. 654pp. 5⅜ x 8½. 40179-0

EXPERIMENTS IN TOPOLOGY, Stephen Barr. Classic, lively explanation of one of the byways of mathematics. Klein bottles, Moebius strips, projective planes, map coloring, problem of the Koenigsberg bridges, much more, described with clarity and wit. 43 figures. 210pp. 5⅜ x 8½. 25933-1

CONFORMAL MAPPING ON RIEMANN SURFACES, Harvey Cohn. Lucid, insightful book presents ideal coverage of subject. 334 exercises make book perfect for self-study. 55 figures. 352pp. 5⅝ x 8¼. 64025-6

THE GEOMETRY OF RENÉ DESCARTES, René Descartes. The great work founded analytical geometry. Original French text, Descartes's own diagrams, together with definitive Smith-Latham translation. 244pp. 5⅜ x 8½. 60068-8

THE THIRTEEN BOOKS OF EUCLID'S ELEMENTS, translated with introduction and commentary by Sir Thomas L. Heath. Definitive edition. Textual and linguistic notes, mathematical analysis. 2,500 years of critical commentary. Unabridged. 1,414pp. 5⅜ x 8½.. Three-vol. set.
Vol. I: 60088-2 Vol. II: 60089-0 Vol. III: 60090-4

GEOMETRY OF COMPLEX NUMBERS, Hans Schwerdtfeger. Illuminating, widely praised book on analytic geometry of circles, the Moebius transformation, and two-dimensional non-Euclidean geometries. 200pp. 5⅝ x 8¼. 63830-8

DIFFERENTIAL GEOMETRY, Heinrich W. Guggenheimer. Local differential geometry as an application of advanced calculus and linear algebra. Curvature, transformation groups, surfaces, more. Exercises. 62 figures. 378pp. 5⅜ x 8½. 63433-7

CURVATURE AND HOMOLOGY: Enlarged Edition, Samuel I. Goldberg. Revised edition examines topology of differentiable manifolds; curvature, homology of Riemannian manifolds; compact Lie groups; complex manifolds; curvature, homology of Kaehler manifolds. New Preface. Four new appendixes. 416pp. 5⅜ x 8½. 40207-X

TOPOLOGY, John G. Hocking and Gail S. Young. Superb one-year course in classical topology. Topological spaces and functions, point-set topology, much more. Examples and problems. Bibliography. Index. 384pp. 5⅝ x 8¼. 65676-4

LECTURES ON CLASSICAL DIFFERENTIAL GEOMETRY, Second Edition, Dirk J. Struik. Excellent brief introduction covers curves, theory of surfaces, fundamental equations, geometry on a surface, conformal mapping, other topics. Problems. 240pp. 5⅜ x 8½. 65609-8

Math–History of

A SHORT ACCOUNT OF THE HISTORY OF MATHEMATICS, W. W. Rouse Ball. One of clearest, most authoritative surveys from the Egyptians and Phoenicians through 19th-century figures such as Grassman, Galois, Riemann. Fourth edition. 522pp. 5⅜ x 8½. 20630-0

THE HISTORY OF THE CALCULUS AND ITS CONCEPTUAL DEVELOPMENT, Carl B. Boyer. Origins in antiquity, medieval contributions, work of Newton, Leibniz, rigorous formulation. Treatment is verbal. 346pp. 5⅜ x 8½. 60509-4

THE HISTORICAL ROOTS OF ELEMENTARY MATHEMATICS, Lucas N. H. Bunt, Phillip S. Jones, and Jack D. Bedient. Fundamental underpinnings of modern arithmetic, algebra, geometry and number systems derived from ancient civilizations. 320pp. 5⅜ x 8½. 25563-8

A HISTORY OF MATHEMATICAL NOTATIONS, Florian Cajori. This classic study notes the first appearance of a mathematical symbol and its origin, the competition it encountered, its spread among writers in different countries, its rise to popularity, its eventual decline or ultimate survival. Original 1929 two-volume edition presented here in one volume. xxviii+820pp. 5⅜ x 8½. 67766-4

GAMES, GODS & GAMBLING: A History of Probability and Statistical Ideas, F. N. David. Episodes from the lives of Galileo, Fermat, Pascal, and others illustrate this fascinating account of the roots of mathematics. Features thought-provoking references to classics, archaeology, biography, poetry. 1962 edition. 304pp. 5⅜ x 8½. (Available in U.S. only.) 40023-9

OF MEN AND NUMBERS: The Story of the Great Mathematicians, Jane Muir. Fascinating accounts of the lives and accomplishments of history's greatest mathematical minds–Pythagoras, Descartes, Euler, Pascal, Cantor, many more. Anecdotal, illuminating. 30 diagrams. Bibliography. 256pp. 5⅜ x 8½. 28973-7

HISTORY OF MATHEMATICS, David E. Smith. Nontechnical survey from ancient Greece and Orient to late 19th century; evolution of arithmetic, geometry, trigonometry, calculating devices, algebra, the calculus. 362 illustrations. 1,355pp. 5⅜ x 8½. Two-vol. set. Vol. I: 20429-4 Vol. II: 20430-8

A CONCISE HISTORY OF MATHEMATICS, Dirk J. Struik. The best brief history of mathematics. Stresses origins and covers every major figure from ancient Near East to 19th century. 41 illustrations. 195pp. 5⅜ x 8½. 60255-9

Physics

OPTICAL RESONANCE AND TWO-LEVEL ATOMS, L. Allen and J. H. Eberly. Clear, comprehensive introduction to basic principles behind all quantum optical resonance phenomena. 53 illustrations. Preface. Index. 256pp. 5⅜ x 8½.   65533-4

ULTRASONIC ABSORPTION: An Introduction to the Theory of Sound Absorption and Dispersion in Gases, Liquids and Solids, A. B. Bhatia. Standard reference in the field provides a clear, systematically organized introductory review of fundamental concepts for advanced graduate students, research workers. Numerous diagrams. Bibliography. 440pp. 5⅜ x 8½.   64917-2

QUANTUM THEORY, David Bohm. This advanced undergraduate-level text presents the quantum theory in terms of qualitative and imaginative concepts, followed by specific applications worked out in mathematical detail. Preface. Index. 655pp. 5⅜ x 8½.   65969-0

ATOMIC PHYSICS (8th edition), Max Born. Nobel laureate's lucid treatment of kinetic theory of gases, elementary particles, nuclear atom, wave-corpuscles, atomic structure and spectral lines, much more. Over 40 appendices, bibliography. 495pp. 5⅜ x 8½.   65984-4

AN INTRODUCTION TO HAMILTONIAN OPTICS, H. A. Buchdahl. Detailed account of the Hamiltonian treatment of aberration theory in geometrical optics. Many classes of optical systems defined in terms of the symmetries they possess. Problems with detailed solutions. 1970 edition. xv + 360pp. 5⅜ x 8½.   67597-1

THIRTY YEARS THAT SHOOK PHYSICS: The Story of Quantum Theory, George Gamow. Lucid, accessible introduction to influential theory of energy and matter. Careful explanations of Dirac's anti-particles, Bohr's model of the atom, much more. 12 plates. Numerous drawings. 240pp. 5⅜ x 8½.   24895-X

ELECTRONIC STRUCTURE AND THE PROPERTIES OF SOLIDS: The Physics of the Chemical Bond, Walter A. Harrison. Innovative text offers basic understanding of the electronic structure of covalent and ionic solids, simple metals, transition metals and their compounds. Problems. 1980 edition. 582pp. 6⅛ x 9¼.
66021-4

HYDRODYNAMIC AND HYDROMAGNETIC STABILITY, S. Chandrasekhar. Lucid examination of the Rayleigh-Benard problem; clear coverage of the theory of instabilities causing convection. 704pp. 5⅜ x 8¼.   64071-X

INVESTIGATIONS ON THE THEORY OF THE BROWNIAN MOVEMENT, Albert Einstein. Five papers (1905–8) investigating dynamics of Brownian motion and evolving elementary theory. Notes by R. Fürth. 122pp. 5⅜ x 8½.   60304-0

THE PHYSICS OF WAVES, William C. Elmore and Mark A. Heald. Unique overview of classical wave theory. Acoustics, optics, electromagnetic radiation, more. Ideal as classroom text or for self-study. Problems. 477pp. 5⅜ x 8½.   64926-1

PHYSICAL PRINCIPLES OF THE QUANTUM THEORY, Werner Heisenberg. Nobel Laureate discusses quantum theory, uncertainty, wave mechanics, work of Dirac, Schroedinger, Compton, Wilson, Einstein, etc. 184pp. 5⅜ x 8½.   60113-7

ATOMIC SPECTRA AND ATOMIC STRUCTURE, Gerhard Herzberg. One of best introductions; especially for specialist in other fields. Treatment is physical rather than mathematical. 80 illustrations. 257pp. 5⅜ x 8½.   60115-3

AN INTRODUCTION TO STATISTICAL THERMODYNAMICS, Terrell L. Hill. Excellent basic text offers wide-ranging coverage of quantum statistical mechanics, systems of interacting molecules, quantum statistics, more. 523pp. 5⅜ x 8½.
65242-4

THEORETICAL PHYSICS, Georg Joos, with Ira M. Freeman. Classic overview covers essential math, mechanics, electromagnetic theory, thermodynamics, quantum mechanics, nuclear physics, other topics. First paperback edition. xxiii + 885pp. 5⅜ x 8½.   65227-0

PROBLEMS AND SOLUTIONS IN QUANTUM CHEMISTRY AND PHYSICS, Charles S. Johnson, Jr. and Lee G. Pedersen. Unusually varied problems, detailed solutions in coverage of quantum mechanics, wave mechanics, angular momentum, molecular spectroscopy, more. 280 problems plus 139 supplementary exercises. 430pp. 6½ x 9¼.   65236-X

THEORETICAL SOLID STATE PHYSICS, Vol. 1: Perfect Lattices in Equilibrium; Vol. II: Non-Equilibrium and Disorder, William Jones and Norman H. March. Monumental reference work covers fundamental theory of equilibrium properties of perfect crystalline solids, non-equilibrium properties, defects and disordered systems. Appendices. Problems. Preface. Diagrams. Index. Bibliography. Total of 1,301pp. 5⅜ x 8½. Two volumes.   Vol. I: 65015-4   Vol. II: 65016-2

A TREATISE ON ELECTRICITY AND MAGNETISM, James Clerk Maxwell. Important foundation work of modern physics. Brings to final form Maxwell's theory of electromagnetism and rigorously derives his general equations of field theory. 1,084pp. 5⅜ x 8½. Two-vol. set.   Vol. I: 60636-8   Vol. II: 60637-6

OPTICKS, Sir Isaac Newton. Newton's own experiments with spectroscopy, colors, lenses, reflection, refraction, etc., in language the layman can follow. Foreword by Albert Einstein. 532pp. 5⅜ x 8½.   60205-2

THEORY OF ELECTROMAGNETIC WAVE PROPAGATION, Charles Herach Papas. Graduate-level study discusses the Maxwell field equations, radiation from wire antennas, the Doppler effect and more. xiii + 244pp. 5⅜ x 8½.   65678-5

INTRODUCTION TO QUANTUM MECHANICS With Applications to Chemistry, Linus Pauling & E. Bright Wilson, Jr. Classic undergraduate text by Nobel Prize winner applies quantum mechanics to chemical and physical problems. Numerous tables and figures enhance the text. Chapter bibliographies. Appendices. Index. 468pp. 5⅜ x 8½.   64871-0

METHODS OF THERMODYNAMICS, Howard Reiss. Outstanding text focuses on physical technique of thermodynamics, typical problem areas of understanding, and significance and use of thermodynamic potential. 1965 edition. 238pp. 5⅜ x 8½. 69445-3

TENSOR ANALYSIS FOR PHYSICISTS, J. A. Schouten. Concise exposition of the mathematical basis of tensor analysis, integrated with well-chosen physical examples of the theory. Exercises. Index. Bibliography. 289pp. 5⅜ x 8½. 65582-2

RELATIVITY IN ILLUSTRATIONS, Jacob T. Schwartz. Clear nontechnical treatment makes relativity more accessible than ever before. Over 60 drawings illustrate concepts more clearly than text alone. Only high school geometry needed. Bibliography. 128pp. 6⅛ x 9¼. 25965-X

THE ELECTROMAGNETIC FIELD, Albert Shadowitz. Comprehensive undergraduate text covers basics of electric and magnetic fields, builds up to electromagnetic theory. Also related topics, including relativity. Over 900 problems. 768pp. 5⅝ x 8¼. 65660-8

GREAT EXPERIMENTS IN PHYSICS: Firsthand Accounts from Galileo to Einstein, edited by Morris H. Shamos. 25 crucial discoveries: Newton's laws of motion, Chadwick's study of the neutron, Hertz on electromagnetic waves, more. Original accounts clearly annotated. 370pp. 5⅜ x 8½. 25346-5

RELATIVITY, THERMODYNAMICS AND COSMOLOGY, Richard C. Tolman. Landmark study extends thermodynamics to special, general relativity; also applications of relativistic mechanics, thermodynamics to cosmological models. 501pp. 5⅜ x 8½. 65383-8

LIGHT SCATTERING BY SMALL PARTICLES, H. C. van de Hulst. Comprehensive treatment including full range of useful approximation methods for researchers in chemistry, meteorology and astronomy. 44 illustrations. 470pp. 5⅜ x 8½. 64228-3

STATISTICAL PHYSICS, Gregory H. Wannier. Classic text combines thermodynamics, statistical mechanics and kinetic theory in one unified presentation of thermal physics. Problems with solutions. Bibliography. 532pp. 5⅜ x 8½. 65401-X

Paperbound unless otherwise indicated. Available at your book dealer, online at **www.doverpublications.com**, or by writing to Dept. GI, Dover Publications, Inc., 31 East 2nd Street, Mineola, NY 11501. For current price information or for free catalogues (please indicate field of interest), write to Dover Publications or log on to **www.doverpublications.com** and see every Dover book in print. Dover publishes more than 500 books each year on science, elementary and advanced mathematics, biology, music, art, literary history, social sciences, and other areas.